Crossing Wires

Crossing Wires

Making Sense of Technology, Transhumanism, and Christian Identity

JOEL OESCH

WIPF & STOCK · Eugene, Oregon

CROSSING WIRES
Making Sense of Technology, Transhumanism, and Christian Identity

Wipf & Stock
An Imprint of Wipf and Stock Publishers
199 W. 8th Ave., Suite 3
Eugene, OR 97401

www.wipfandstock.com

PAPERBACK ISBN: 978-1-7252-8732-7
HARDCOVER ISBN: 978-1-7252-8733-4
EBOOK ISBN: 978-1-7252-8734-1

Manufactured in the U.S.A. 11/18/20

Contents

Introduction

The Beginning of the End

Eagle Eye

IMAGINE THAT YOU GO to an eye doctor for your annual check-up. Annoyed by your ever-decaying range of sight, you move numbly from one test to another, all to confirm what you have suspected for some time now: you need stronger lenses. Your doctor hits you with some news, but it's not quite the news you expected.

She cheerfully says, "Your eyesight is, in fact, getting worse. That's the bad news. The good news is that we live in truly remarkable times. In conjunction with *Eyeborg Industries*, our office is running a beta program that allows us to offer one patient a special opportunity." Caught slightly off guard, you are nevertheless interested in what your doctor has to say.

She continues, "The world of robotics has advanced to the point that an artificially-built eye can now be inserted into your eye socket and be directly connected to your optic nerve, taking the place of your own failing biological eye. This new eyeball would give you perfect vision at a hundred yards." For a moment, you recoil at the thought of a robotic eye, since the only image entering your head is a laser-eyed killer robot from some 80s sci-fi flick. Before you respond, however, the doctor charges on with the benefits . . .

"Now I know what you're thinking. One, ouch. Two, weird-looking. Three, expensive. Four, well, I'm sure you're thinking of a four. Give me five minutes of your time so you can make a well-informed decision. The process is neither painful nor particularly dangerous. You can come into our specialized operating room and be out with your new eye in no more than three hours. In addition, I'm pleased to say that a robotic eye is impossible to distinguish from a normally functioning human eye—they are identical in every way, both in looks and in movement."

You're not sure how to respond, but you've now committed yourself to the duration of this presentation, so you say nothing. The ophthalmologist concludes with this zinger, "I can offer you this technology for free, simply because *Eyeborg Industries* wants some initial good press before offering the product to the greater population."

How would you react?

If all of the scientific data stood up to your scrutiny, would you be willing—perhaps even excited—to receive a robotic eye? This little scenario stands on its own as an interesting thought experiment, but what would you say if the doctor's proposition grew in capacity and in *controversy*? Let's continue. The doctor tells you excitedly that the eye implant has just received a significant upgrade. It now includes the ability to see at a thousand yards with perfect clarity and includes the ability to see objects in x-ray, zoom, and night vision! The experiment now has taken on a new dimension. At first, you were under the impression that the robotic eye would simply allow you to recover a range of human eyesight that is at least plausible (though certainly enhanced for most of us). Now, you would have distinctly super-human abilities internal to your own body. If you were in the position of the patient, at what point in this little narrative would you say, "Thanks, but no thanks"? Or, perhaps, nothing concerns you enough to say "no thanks" in the first place!

Would your decision change if the doctor said, "All this can be yours, but if you get the surgery done, I have to offer the advanced version, including all the bells and whistles, to everyone. Your neighbors. Your local police force. Your nosy neighbor Steve"?

When I tell this thought experiment to my students (a conversation starter I've affectionately named, "Eagle Eye"), I usually find their responses to be quite enthusiastic in the beginning, turning darker and more critical as the doctor's proposal gains in force. It's the *why* that interests me, not the how or what. I ask the students, "Why were you okay with a robotic eyeball under *these* conditions, but not *those*? Why did you like perfect sight at a hundred yards, but recoil at the thought of x-ray vision?"

Two types of negative responses generally rise to the surface. The first response is primarily individualistic in nature. It typically sounds something like, "I know myself too well to use these abilities in a positive way over any long stretch of time. I might use them properly for a few days or weeks, but the temptation to abuse them is simply too great." In particular, technologies associated with extending the capacities of our eyes have often led to a well-founded concern about the invasion of privacy. We would see too much. It's the same gut feeling we would get if we were to visit a friend's high rise loft to find a telescope resting near the window. We would simultaneously ask ourselves if our friend was actually a creeper, and then in the very next thought, we'd want to try it out!

As a theologian, I find that there is much to admire about the honesty present in such a response. Humans are particularly gifted at one thing: making a grand mess of things. Less common is the ability to recognize our personal capacity to contribute to the mess. Classically, Christianity has termed our human failure as "sin" or "depravity," and history has a way of confirming this condition as an ever-present challenge. Humans build glorious colosseums then fill them with idols and gladiators. They create information networks that boggle the mind then quickly use them to distribute fake news or pornography. What new horrors could our species invent with 1000/20 vision?

The second response to the above thought experiment is more communal, not in the sense that the community shapes the answer, but rather that the answer speaks to the way we think about ourselves *as a species*. Such a response might sound something like this: "Giving me the ability to perform so far outside of the normal spectrum of human behavior makes me wonder if I'm actually changing what it means to be human in the first place. It feels like I'm becoming less human when I replace parts of my body." Notice that the concern here is less about the potential for widespread abuse and more about the radical transformative power that technology can generate *within human nature*. A human is a human, right up to the point where a person does something decidedly inhumane or, better yet, *un*-human. Is it easier to think of Charles Manson as a fellow man or as a monster? Considering the widespread influence of today's digital technologies, it makes perfect sense for us to wonder whether these forces are so powerful that they change us into something *different*. What exactly makes humans, human, after all?

A Roadmap to the End of Humanity

Welcome to a discussion that is just now beginning to pop up in news feeds and percolate around coffee shops. It's much more than a discussion about the newest gadget or smartphone app; it's a high stakes game about human identity, potential, and the drive to be happy. We're all involved because each one of us uses digital technologies. We're all involved because it is part of the human experience to ask the big questions: Why am I here? What kind of person do I want to be? How do I understand my role within the context of my community? No other species asks these questions, and no other species uses tools to help them get the proper answers. We use digital technologies to improve nearly every portion of our lives, from applications in social networking to shopping to banking to information-sharing. Humanity alone has the capacity to reflect deeply about those tools and what they are doing to us. As the famous saying goes, "We shape our tools, and thereafter our tools shape us."[1]

This is a discussion about the end of humanity as we know it. It is also a discussion about what happens *after* the end of humanity. No longer the domain of clever futurists or science fiction films, the developing technologies of today are calling into question what it means to be human going forward. Many people are calling for a total system upgrade where our biological organs are increasingly enhanced, even replaced, by smartly applied technologies. It is time, they say, to let this current version of man and woman to fall into the flaming brush pile of history so that a new phoenix can rise from the ashes. Call it Humanity 2.0. Supporters argue that embracing technology is the best (perhaps only) way forward; the path to happier, healthier communities comes by welcoming innovation to a degree that makes us *transhuman*—somewhere on the path between current humanity and a glorious future of modified, advanced beings that make up "posthumanity," a future so bright, some argue, that we really can't call these augmented people "human" anymore. Those willing to embrace this merge with technology reap the benefits of being exponentially smarter, stronger, and happier than those who plod along as "normal" people. For transhumanists, the end of humanity doesn't come on the back of atomic

1. Mistakenly attributed to sociologist Marshall McLuhan, this phrase was actually penned by his friend, Father John Culkin, a professor at Fordham University. The quote surfaced in "A Schoolman's Guide to Marshall McLuhan" in *The Saturday Review*. It has been modified and used in a variety of contexts since then, though it is no stretch to think that the central idea behind the quote is McLuhan's.

mushroom clouds or melted ice caps. Rather, it's just the next stage of evolution hurried along by the light-speed evolution of the microprocessor.

From the outset of this discussion, let's keep something important in mind. While humanity does not yet hold the capacity to cure all diseases or manufacture limitless happiness through gene therapies or chemical enhancements, the philosophy of a trans- or post-human future is already here. Titans in the tech industry are busy transforming the world as you know it, and the effects are being felt across every facet of culture, from entertainment to politics, from health to law, from psychology to religion. At this very moment, governments are considering laws that give moral status to robots. In the past, only living, breathing, organic beings have been given this dignity: orcas, Labrador puppies, bald eagles, black rhinos, and the sort. Now, legislators are deliberating whether or not our *machines* require protection under the law as personal beings![2] Ever want to take a baseball bat to your office printer? Well, that just got a bit more complicated.

These industries are asking (and answering) crucial questions that, up to this point, have been left largely in the hands of scientists and philosophers. Questions like, is human nature changeable? Can we shape our future destiny by altering the physical bodies we have been using all this time? Does our society have a moral obligation to encourage forms of research that increasingly extend our life spans? The church can no longer stand on the sidelines because these complicated realities are upon us. It must enter the fray and engage the questions that may or may not have clear biblical answers. Can a Christian in good conscience receive a cochlear implant or Lasik eye surgery? What about gene therapies? Memory implants? These are thoughtful questions to entertain, no doubt, but they were not on the radar of the first-century church in Jerusalem, so we'll have to be careful when and how we apply our Christian critique.

The following chapters will provide you with a primer on the world of digital technologies and the transhumanist philosophy that carries them forward, both with its possibilities and its consequences. Many of my observations will be borne out of a Christian theological perspective, but not entirely. It is crucial that we not only view the brave new world of human enhancement with a strong biblical foundation but also employ

2. Members of the European Parliament, in one example, recently recommended a set of comprehensive rules for person-robot interactions. They are considering legislation that would recognize certain robots as "electronic persons" for the purpose of giving them some form of moral status. See Wakefield, "MEPs vote on robots' legal status." The article includes a link to a draft of the legislation.

straightforward reasoning disconnected from religious commitments. A non-Christian—even an atheist—has many reasons to be hesitant about the supposed utopian future many transhumanists are promising. Conversely, Christians and non-Christians can stand shoulder to shoulder and joyfully participate in a life made better by technological advances. We will address these arguments in due time.

If the church were to ignore the substantial assumptions present in Transhumanism as fanciful or unrealistic, this book should serve as a warning. Transhumanism, in many ways, is already here. I will be spending the middle portion of this book showing you the remarkable ways digital techs are being used today to advance transhumanist ideals. However, a thoughtful approach by the church—and by that, I mean *you*—can lead the way by providing a helpful critique of the movement without resorting to extremes on either side: blind love of technology on one side or outright rejection of anything with a WiFi connection on the other.

The church, in the broad sense, is filled with Christians (shocker). Christians are charged with discipling, baptizing, and teaching all nations in the name of Christ (Matt 28:18–20), and therefore must be willing to encounter a culture that knows little or nothing about *the ways of Christ.* If Jesus is the Lord of all, which all Christians proclaim, the responsible Christian must navigate how to apply sound biblically based theology to a movement that largely sees Christianity (and to a lesser extent, most other religions) as the enemy of progress. I am not going to gloss over the deep-rooted philosophical differences between Transhumanism and confessional Christianity; they are substantial and important. However, I believe there is a way forward that will allow believers to engage this cultural phenomenon with humility, freedom, conversation, and grace. As the military-inspired proverb goes, an army that is wary of the ambush is never caught unawares. The first step, then, is to lay bare the terrain and find out where precisely transhumanist thought can be a threat and where it can be highly beneficial and worthy of our cautious support.

A word of warning: the terrain can get a little weird. In good time, I'm going to introduce some cultural phenomena that you don't often talk about in respectable Christian circles. We'll encounter sex robots and gender-bending, memory implants and mind uploads, hedonism and head-freezing (yes, you read that right). It won't be a trip for the faint of heart, so be strong and courageous. The destinies of our communities hang in the balance, and our children will be looking to us for guidance. It's okay to experience a little gallows humor, where the craziness of the world evokes simultaneous fits of laughter and shouts of, "That's *insane*!" This is not your grandparents' world—it requires *your* discernment, *your* commitment, and *your* community to bring it into proper perspective.

With this in mind, the Christian solution to the problems of the world remains the same as it has professed for the last two-thousand years. The solution is Christ. Now, you may not need this book to tell you a punchline that has been the answer to every Sunday school question since the beginning of time. What we may need, however, is a fresh look at the theological resources embedded in God's story that can offer a particularly poignant and discerning response to Transhumanism.

The question that flows through this book as an unmistakable undercurrent is, "What makes humans, human?" We'll come back to this time and time again. The only way we will get to the destination of true human flourishing is to nail this question.[3] But this will take some time, so have a little patience. I will slowly but steadily introduce passages from Scripture as well as theological concepts to shore up my argument. Eventually, I will offer up a workable definition of humanity, arguing that a Trinitarian perspective on personhood is a helpful way to stimulate discernment and wisdom in an age of instant answers.

The Role of the Believer

What about you, the Christian man or woman? Why should you care? Why invest your time and effort to read a book about robots and virtual reality?

I'm inviting you on this journey because, whether you like it or not, the internet has changed you . . . and it's not going to stop. If we get our approach to technology wrong by being passive and unreflective, the Digital Age[4] will transform our communities, our children, and our futures into data-driven digi-gods that lack empathy, let alone the Christian fruits of the Spirit that foster relationships of love and belonging. If, however, we view technology with clear eyes and humble hearts, the possibilities to serve

3. The term, "human flourishing," is classically attributed to Aristotle and his use of the term, *eudaimonia*. It largely refers to the good life, a life marked by the use of reason and guided by the chief virtues.

4. This term refers to the era of time from around 1970 to the present, where our economy has largely developed on the back of the microprocessor, away from more traditional industrial sources of economic and cultural growth. It has often been called the Age of Information.

one another under the banner of hope, faith, and love are virtually (pun intended) endless.

Christians are a diverse bunch; each one is a unique child of God. As you read this book, I encourage you to bring the principles you glean back to your specific context, finding ways to apply what you have learned in your homes and local communities. Are there insights that you can apply to Christian life, both in and out of the church context, which might help you and your tribe consider technology and Transhumanism in a responsible, godly manner? My sincere belief is that this endeavor is not only possible, but imperative. The individual Christian does not have the luxury of waiting for her church body to issue a position paper on Transhumanism and then defer to its conclusions. Outsourcing our thinking on such matters creates intellectually lazy Christians. The more familiar the average believer is with these concepts, the more often they are able to engage the broader culture with confidence. If enough Christians have respectful, winsome conversations with their neighbors, the biblical view of human flourishing will be primed for an epic comeback. Grace, freedom, and the cross will yet again prove to be the most valuable resources in our arsenal. Make no mistake: this view will often fall into direct confrontation with the transhumanist movement, so buckle up.

Much like Christian theology, the topics of technology and Transhumanism will require from time to time the use of some painful terminology. Elaborate concepts will be introduced here. Don't give up when you confront a difficult section! I will do my best to lay out the central features of each piece of the puzzle in a way that you can understand, though some concepts might need some extended reflection. For this reason, I strongly suggest reading this book one chapter at a time. Don't bulldoze through it and miss the finer points in the name of speed. Read a chapter. Think about it. Talk about it. Write down some thoughts or a few side questions. Rinse and repeat. This book will work best if you share the experience with another kindred spirit, a spouse, or your Bible study group. The questions that are raised here will only scratch the surface of the broader conversation our churches need to be having with the culture at large. Your *community* is required for any long-term benefit, so enlist them! Read this together and see what blooms from your discussions, even if those discussions get a little heated!

Bringing out the pitchforks and storming your local Apple superstore is not the answer prescribed in this book. I am calling for the people of

God to be what they already are: bearers of Christ in a broken world. Use this book to invite both Christians and non-believers into a conversation. Transhumanism has the potential to affect nearly every aspect of our lives, regardless of our demographics. Look what effect the online social network craze has on our community right at this moment, and remember that Facebook has only been around since the mid-2000s. The earlier these conversations start, the more time you (and your neighbor) have to absorb, then critique, the deep and lasting implications of the Digital Age for our society and our children. Only good things can come from communities that talk with one another about difficult and complex issues. I encourage you to take up the mantle of responsible, respectful conversation in your families, your neighborhoods, your churches, and your civic groups, even—perhaps, *especially*—when you do not see eye to eye with your neighbor! This great task will not be easy, for the roadmap through the digital haze is fraught with dangers on all sides. In the words of the immortal Axl Rose: "Welcome to the jungle."

Discussion Questions

1) Before reading this chapter, had you heard of the term "Transhumanism" before? In what context? Can you define it?
2) What was your response to the Eagle Eye experiment mentioned above? Were you comfortable with all of the "modifications," or were you hesitant at certain junctures? Why?
3) If you had to place your present position on digital technology along a spectrum, how would you do it? Would you consider yourself more techno-friendly or techno-critical? How would you describe your church's position?

used to perish each day. We are heirs of the [illegible] spoken word. [illegible] Christianity and [illegible] the [illegible] conversation. Transhumanism [illegible] every aspect of our lives [illegible] Look [illegible] of the [illegible] [partly illegible]

[illegible] our [illegible] right at the [illegible] [partly illegible]

[illegible] and [illegible] have to absorb the [illegible] implications [illegible] [partly illegible]

[illegible] encourage you to [illegible] conversation in your families [illegible] [partly illegible]

[illegible] Welcome [illegible] [partly illegible]

Discussion Questions

1. [illegible] have you heard of the term "transhumanism" [illegible] How would you define it? [partly illegible]

2. What has been your experience [illegible] [partly illegible]

3. [illegible] present [illegible] technology [illegible] church's position? [partly illegible]

Chapter 1

A Brief Introduction to Transhumanism

Neutral Tools?

IMAGINE WAKING UP IN the morning to a feeling of overwhelming contentedness. Almost instantly, your mind is clear, focused, and creative. You live in a beautiful home filled with art from your own hand, and as you stroll to the kitchen, you hear transcendent music of your own composition. You are a picture of perfect health and everyone knows it, though you do get curious stares from time to time because of your custom violet-colored eyes. Taking a quick bite to eat, you are filled with an indescribable hope for the day's possibilities, your mind instantly and perfectly cataloguing the day's schedule: meditation, poetry, and a hike in the nearby wilderness. Even your very muscles bristle with excitement, knowing that today—your three-hundredth birthday—is just the beginning of the rest of your life.

The next twenty years of digital and robotic technologies may usher in this sort of future, a world of almost unbelievable opportunity. Or, perhaps, the future is filled with information and connectivity and yet emptied of empathy and genuine community. Thoughtful people can recognize that technologies have the potential to be put to good or bad uses. Often, it's not that easy to discern which is which, as one man's paradise is another man's prison. Because people are generally uncomfortable with calling *things* (rather than *acts*) moral or immoral, the tendency is to draw focus away from the tool and toward the person who uses it, and rightly so. My power drill isn't evil (at least, I think it's not). I would have to use it toward an evil end, something that is entirely in my power.

Nevertheless, we want to be careful not to press this too far by saying a seemingly benign phrase like, "All technology is inherently neutral. How you use it makes an innovation good or bad." I encourage you to consider that not all technologies are equal. Some tools, even when operated as intended, change the user in some fundamental ways. A husband who brings a smartphone on vacation will have a distinctly different experience than his wife who purposely left hers at home. He is more likely to be set at edge, more likely to be a slave to an arbitrary itinerary, and certainly more likely to drive his wife crazy than *vice versa*. I no longer agree that tools are fundamentally neutral. It's more of a half-truth. Digital technologies, in particular, shape a person at a far more foundational level than we realize, and as these technologies increase in power and complexity—which they will—the transformations that take place in the user will increase with corresponding power and complexity. Transhumanism favors the ongoing experimentation of life-altering technologies toward the noble ends of life extension and general well-being. The tools required to achieve these audacious goals must be thoroughly scrutinized since the explicit goal of Transhumanism is to change humanity into something else!

Before we confront Transhumanism on its own terms, it's valuable to note that this is not some trite philosophy existing in the shadows of a few think tanks or rogue academic departments. This is a "now" issue, not a "somewhere down the line" issue. At this very moment, brave futurists surgically insert magnets into their fingertips, effectively giving them a sixth sense. Driverless cars are quickly becoming a ho-hum sight in several American cities. Robots are being used for geriatric care in Japan and as receptionists in Singapore. People are naturally excited about the possibilities in health medicine; they should be. The pace of technological progress in the past decade alone is staggering. Transhumanism simply takes this enthusiasm toward innovation and drives it to its logical conclusion. If the normal human predicament is filled with difficulty, pain, and limitation, why not harness the power of technology to make these hardships obsolete? Why *can't* we live with health, happiness, and power? If you have difficulty summoning an objection to this, then perhaps you are already thinking like a Transhumanist! Now, let's take a look at what Transhumanists believe, where they want to go, and how they plan to get there.

Putting the "Trans" in Transhumanism

Putting the prefix "trans-" in front of anything these days seems to be the politically correct thing to do. The staggering amount of public attention the media paid to Caitlyn Jenner has brought the issue of transgenderism to the forefront of gender studies and cultural life in America. Trans-racialism has hit the scene, as well: a term that suggests that biological race and racial identities are two separate concepts (the latter as a social construct) but no longer inseparably connected. For example, a young woman could have two Caucasian birth parents but still consider herself "black." Trans- is just another way to say fluid or moving.

Transhumanism (shorthand, "H+") is the belief that human nature is not a fixed concept; in fact, humanity can use technology and applied reason to break free from the shackles imposed by its organic bodies to live longer, better lives. Said another way, H+ is the belief that human nature is an ever-changing thing, and since it is, humans can and should change it according to their own personal desires using every means at their disposal. If you want robotic arms and legs to lift more and run faster, you have the right to augment and upgrade your body to any degree you deem desirable. If you want a better brain with access to the internet at the speed of thought, H+ will be the way to make that dream a reality. No adjustment is too radical, so long as you don't stand in the way of another person's freedom to alter themselves. *Trans*human, for true believers, is just another name for *super*human.

The body, in this view, does not contribute anything to our essential human identity. It is the mind that makes us who we are. As you can see, we are asking the "what makes humans, human" question. Transhumanists place an overwhelming emphasis on the mind as the necessary feature of personhood. They lament the fact that our bodies are severely limited by its natural restrictions such as cranial capacity (brain size), muscle ability, eyesight and hearing limitations, and so on. These are just evolution's arbitrary restrictions. Why not defy evolution, they ask, and give yourself a surgical advantage—say, another eye on the back of your head or a skin graft that allow you to generate photosynthesis?[1] Some transhumanists have been quick to say that this desire to "upgrade" should not be confused

1. Zoltan Istvan, the 2018 Libertarian candidate for governor in California and a self-proclaimed transhumanist, referenced these particular forms of body modification at a lecture he delivered to the Crosswise Institute at Concordia University Irvine in June 2017.

with a hatred of one's body. Max More, a leading transhumanist author and founder of the Extropy Institute,[2] writes,

> In reality, transhumanism doesn't find the biological human body disgusting or frightening. It does find it to be a marvelous yet flawed piece of engineering. It could hardly be otherwise, given that it was designed by a blind watchmaker, as Richard Dawkins put it. True transhumanism *does* seek to enable each of us to alter and improve (by our own standards) the human body and champions morphological freedom. Rather than denying the body, transhumanists typically want to choose its form and be able to inhabit different bodies, including virtual ones.[3]

There is a lot here to sift through. It's relatively easy to note the evolutionary worldview that H+ must emerge from, and because of that, it shouldn't surprise you to find that H+ is comfortable with pressing forward into new forms of human life. After all, enhanced humanity is just the next stage in evolution. In addition, the theme of freedom is used to justify a person's bodily "adjustments." The ability to choose one's own form—even choosing what *kind* of bodies—is a supreme value. The most ambitious transhumanists are researching ways to copy one's consciousness on to a computer system so that a person could escape his body altogether. As we will see a bit later on, this type of freedom will require some careful consideration.

To be sure, the vast majority of transhumanists are atheists, at least in the traditional sense. They would probably be best categorized as humanists, those who believe in the creative capacity and agency of humanity to form a better world today. Christians, too, maintain a healthy concern for their communities, but to think of the body as a wholly malleable concept, a big pile of play-doh to be molded, seems at odds with how man and woman are understood in the Old and New Testaments. One would think that H+, at first blush, would run counter to the Christian worldview, as it blurs the distinction between Creator and creature. Yet this is not entirely accurate, for reasons we shall soon see. Some Christian thinkers have lent some cautious support the H+ movement, viewing it as a philosophy that the Christian faith already largely embraces.[4] After all, they argue, don't Christians

2. Extropy, as defined by transhumanist Max More, is understood as "the extent of a living or organizational system's intelligence, functional order, vitality, and capacity and drive for improvement." More, "Philosophy of Transhumanism," 5.

3. More, "Philosophy of Transhumanism," 15.

4. Redding, "Christian Transhumanism Is the Next Reformation."

believe that they are moving from the old world to a new creation *à la* 2 Cor 5:17? From life to eternal life? From sinful creature to sanctified saint? Christians have long taught that the current condition of the body is not in its final state at the resurrection of the dead where disease and death no longer hold sway. Could the Christian understanding of eternal life be fit into a transhumanist's call for radical drug and gene therapies? We'll get there soon enough, but for now, keep an open mind as you examine the claims of Transhumanism—you might find yourself in agreement with many of its core principles.

Garden variety Transhumanism is committed to a form of progress (in the form of self-enhancement) that is entirely human-driven. No transcendent or external force (i.e., God) wills humanity into a glorious future; those bright days must be achieved by the few who are brave enough to defy their own nature. And doing such requires a journey of sorts.

Return to the term, "transhumanism." The prefix "trans-" has more to tell us. It implies moving from one position to another—to go across something. If I were to buy a ticket for a *trans*-continental flight, I could be sure that I wouldn't just be traveling from Los Angeles to San Diego. The "trans-" prefix demands that I cross some sort of boundary, that I cover a distance from one space to another. In the case of Transhumanism, the journey takes us from human to post-human. But this brings up another issue. Transhumanism is not an unstoppable force; we can choose or reject its philosophy. The implication is that whatever destination this "trans-" trip takes us, *it's a place where we actually want to go*—that the *there* is worth pursuing in the first place. Returning to the above example, I might choose to take a trans-continental flight from Boston to Los Angeles for many reasons: My family lives in California, the weather is warmer in January, and I like fish tacos. The destination is worth the great difficulty and finances involved in traveling. The *there*, in this case, is worth the *trans-*.

Any journey is worth it insofar as the destination is worthy of pursuit. So, we must ask right at the outset, what does it mean to arrive at a post-human destination? "Human" is a straightforward term to understand. *Post*-human, by contrast, is a rather murky affair. I will be returning to this concept throughout the book to explore whether or not post-humanity is something worth pursuing, even if Transhumanism wins the day and all of its hopes for humanity are realized. For their part, many transhumanists are wary of destination talk, rejecting naïve visions of a utopia that lies somewhere in the future. They are more likely to speak of perpetual improvement. This language suggests that humanity will never necessarily *arrive anywhere*—just that the journey is of chief importance. Again, you may be sensing the evolutionary foundation of Transhumanism. Development, improvement, progress. With these principles, the holy grail of human flourishing can be attained. Right?

The Three Supers

The most straightforward way to describe the purposes of H+ is to introduce you to the Three Supers: super-longevity, super-intelligence, and super-well-being. We will do a quick survey of the scientific advancement in these areas, and then at the end of the chapter, we will look at these features through a biblical perspective. For the time being, let's hear what Transhumanism and some of her leading proponents have to say.

Super-Longevity

All the aims of Transhumanism are spokes connected to one central hub: life extension. "Super-longevity," as it is sometimes called, is the scientific push to find ways in which humanity can beat back its own death. Traditionally, we think of death as something that just . . . happens. It's inevitable, painful, and final. Not to be trite, but the evidence here is pretty one-sided. You're going to die—at least, that's what people have assumed for the past few millennia. Technological advances, however, are expanding exponentially at such a rate that many scientists are beginning to believe that death is no longer something we just have to accept. In the near future, they argue, our children will be living well past a hundred—even two-hundred—years. If death is a matter of the body decaying beyond repair, then the goal is simple: locate the places where damage is happening and arrest it . . . even reverse it.

Super-longevity moves the discussion of death away from the body (i.e., when will my body give out?) and toward the will (i.e., when do I want to die?). Sure, some types of death are unforeseeable and unpreventable—a bad car accident, for instance, may hopelessly wreck the body—but in the case of normal, low-risk living, transhumanists argue that super-longevity is not only possible, but inevitable. This is an extraordinary claim. Think of all the parts of your life that would be affected if you were able to extend your life another hundred years. If this included an extension of child-bearing years, would you wait longer to have kids?[5] Would you have them at all? What about "risky" behaviors like driving a car on a freeway or skydiving . . . does the prospect of a super long life force you to rethink your daily activities? What would be the transaction?

Transhumanists start with a profoundly simple question, "Why do we treat death as inevitable?" For a moment, place aside any temptation to respond with Christian theology and consider this question at length.

Aubrey de Grey is a leading researcher in the field of life extension, making a name for himself by becoming the public face of the super-longevity movement. His TED talks have millions of views. He believes that it is humanity's moral responsibility to combat the effects of aging. So much so, that he calls the fight for super-longevity as "the single most urgent imperative for humanity."[6] His case is simple:

Death ends us.

The end of any of us is a tragedy.

Therefore, we should do everything we can to keep living.

Opposing this argument looks like folly. People avoid death every day. We assume death is bad; that's why we take care of our bodies! Even conservative Christians would have to concede the simple sense of this point. In 1 Cor 15, Paul calls death the final enemy to be defeated! Super-longevity's basic mission to extend life looks to be on solid moral ground even when viewed from a religious perspective.

For transhumanists, the problem is two-fold: aging and death. Aging is a problem because it is the cause of our end, plain and simple. As we age, our bodies pick up wear and tear, leading to a variety of harmful mutations and general cellular decay. If this cellular damage passes a certain threshold, the body shuts down and dies. Anti-aging scientists, then, are hard at work to identify the reasons for the cellular damage and find ways to minimize, stop, or altogether reverse the process. After all, the goal is not simply to live long, but to live in maximum health.

Death is a problem because we do not choose it. It chooses us. Because we have no say in the matter, our sense of freedom is violated. Transhumanists consider an individual's freedom the highest value, and therefore, they seek to preserve one's ability to live (and die) in the time and manner

5. Marriage itself would undoubtedly transform. Would you be willing to utter the words, "'Til death do us part" under the expectation that your future spouse would live for five hundred years or longer?

6. De Grey, "Curate's Egg," 215.

of their own choosing. This freedom can be realized in a person's physical appearance (e.g., I can look how I want, dress how I want, etc.), the way a person thinks, or the way a person acts. Death is the enemy of freedom. But once death is overcome by science (hypothetically), this antipathy no longer exists and the power of free choice is returned to the individual. If someone has lived, hypothetically, for six-hundred years and is tired of endless political cycles and car insurance commercials, he can *choose* to die if he so pleases. This is the type of death that transhumanists can tolerate, because *his will* generated the decision to die, not nature. But then again, couldn't this be considered another form of suicide?

Rolling all these points together, transhumanists are utterly committed to longer lives. When they see that 100,000 people a day die by aging, they consider this a needless and senseless tragedy. Why can't governments devote every resource to preventing such a catastrophe? Maybe they have a point, given that government agencies are notorious at spending money on frivolous pursuits; perhaps we should demand disciplined government spending to divert any and all financial resources to life-extension research. The potential benefits are almost impossible to calculate. For example, if technological progress can add more than a year of life expectancy *per year*, a term known as the "longevity escape velocity," then humanity has left behind the grip of death forever. Boom. Super-longevity. All that is required, proponents argue, is brute force government funding and human ingenuity.

But would super-longevity cause as many problems as it solves? Virtues such as sacrifice, charity, and courage oftentimes require the fear of death in order to be powerful catalysts of character. Only by *facing death* (or another steep consequence) does bravery exist at all; without it, the term is meaningless and the virtue cannot be cultivated. With nothing serious at stake, would humanity descend into a shallow lifestyle of appetites and urges? After all, virtue may no longer be demanded of them. Francis Fukuyama, one of the leading critics of the transhumanist movement, offers this food for thought:

> Death may come to be to be seen not as a natural and inevitable aspect of life, but a preventable evil like polio or the measles. If so, then accepting death will appear to be a foolish choice, not something to be faced with dignity or nobility. Will people still be willing to sacrifice their lives for others, when their lives could potentially stretch out ahead of them indefinitely, or condone the sacrifice of the lives of others? Will they cling desperately to the life

> that biotechnology offers? Or might the prospect of an unendingly empty life appear simply unbearable?[7]

At first glance, extreme life-extension appears to be fairly untouchable; it is a virtuous goal. Yet critics like Fukuyama are wondering whether *unlimited* life damages or delays the development of character formation, the very disciplines that make for wise, honorable, and cultivated citizens *as they relate to the community at large*. What is good for the individual is not necessarily good for humanity. In a strange irony, death is in part what makes life so precious, so valuable.

Living longer and living forever are two different things, would you agree? It could be argued that the Bible supports both of these ambitions. Should Christians, in particular, get nervous about the super-longevity that H+ seeks to make reality?

Super-Intelligence

In some ways, the second "Super" is the beginning of the end. The end game for Transhumanism is post-humanity, a state in which human knowledge and enlightenment reach profound levels. The only way this result can be achieved is to create a situation where society acquires vastly higher degrees of intelligence than it currently has. So, how does a culture increase its intelligence—not by a little, but by a lot?

Humanity has always pursued knowledge. In an evolutionary worldview, early humans needed to acquire knowledge for survival—passing on crucial information to their offspring (e.g., where the best fishing holes were or how to make warm clothing). In a Christian worldview, much is the same: we've wanted knowledge from the very beginning. The earliest temptation of Adam and Eve took root in the desire for more knowledge: "So when the woman saw that the tree was good for food, and that it was a delight to the eyes, and that the tree was *to be desired to make one wise*, she took of its fruit and ate" (Gen 3:6, my emphasis). People will do just about anything to understand their world better and their place within it—even chop their way out of paradise, if necessary. That fact alone should give us pause while we consider the allure of knowledge accumulation.

Super-intelligence is a tempting prospect. Just as humanity desires to live longer lives, so they also have an unlimited thirst to know more

7. Fukuyama, *Our Posthuman Future*, 71.

about . . . *more*. If one lives longer and longer without growing in intelligence simultaneously, the overall improvement to the human condition seems minimal. As a topic of transhumanist thought, super-intelligence usually comes in two separate packages, though both fall under a broad definition offered by transhumanist philosopher Nick Bostrom. The term refers to "intellects that greatly outperform the best current human minds across many very general cognitive domains."[8] In English, this essentially means super-intelligence is reached whenever something thinks far faster or smarter than a normal person across a wide variety of tasks. Super-intelligence can come from a computer, therefore, if it can perform with such power and flexibility. Or, it can come from an augmented human, e.g., a man has a computer chip installed on or in his brain to artificially speed up his thought.

Let's break down the two packages of super-intelligence I've just identified. First, super-intelligence can refer to the future moment in time when computers process at the speed of a human brain. It's a type of historical marker. At some point in the future, you will read a headline that says, "The Age of Super-Intelligence." This will mean that computers have finally developed the ability to process information well past the speed of a normal human being. This is more than just brute force computing. When it comes to narrow tasks like mathematics, computers have been walloping us for some time. This is part of the reason we have home computers in the first place! Some of you remember the first time IBM's *Deep Blue* beat chess Grandmaster Garry Kasparov in 1997. At that time, the computer could calculate about two-hundred-million moves *per se*cond. Yet this doesn't qualify as super-intelligence because the computer itself is restricted to one skill—play chess, and that's it.[9] To be honest, *Deep Blue* cannot actually play chess either! It simply processes information and probabilities based on its programmers' commands and the inputted information gathered from previous matches—then spits out the most advantageous move. The computer knows absolutely nothing on its own; it simply does not think for itself—in singular tasks *or* in general intelligence.[10] Ask *Deep Blue* to order a cup of coffee at Starbucks, and I'm afraid you'd be out of luck.

8. Bostrom, *Superintelligence*, 52.

9. Similarly, Grandmaster Lee Sedol was recently defeated by a computer program called *AlphaGo* in a highly publicized tournament in 2016.

10. A brief analysis of the *Deep Blue* program (as it relates to conscious thought) can be found in David Bentley Hart's work. This text can be difficult at times, but it is well worth the effort. Hart, *Experience of God*, 212–25.

For super-intelligence to be attained, *general* intelligence must be achieved. This is the type of broad-based intelligence that people use every day without thinking about it: the strange physics that comes with tying a shoe, the ability to read your boss's facial expressions, the creativity that comes with dancing the tango. These are all features of general intelligence, and robots/computers are nowhere close to this . . . yet.

You could say that computer-based super-intelligence has arrived when either: (1) computers can do all the intellectual tasks a human can do, but do it much faster; or (2) computers think at the same speed as a human but do so in a qualitatively smarter way.[11] Just as humans are qualitatively smarter than horses or dogs, so super-intelligent computers are smarter than humans. We are not talking a little bit of difference here; it is a quantum leap.

The second way in which super-intelligence is understood involves humans more directly. The term can refer to technologies embedded directly in human bodies that dramatically enhance an individual's intelligence. For instance, technologies will emerge in the next ten to twenty years that may substantially increase memory, perhaps even facilitate photographic memory. To use a theoretical example, imagine an operation where a surgeon attaches a neural chip directly to your brain, allowing you to instantly and perfectly bring to mind anything in the Library of Congress, as well as the entire online content of Wikipedia. Certainly that would qualify as super-intelligence, insofar as the recipient has the ability to interpret all that information. We have moved into the realm of the "augmented" person, where the lines between mind and machine quickly become fuzzy. It shouldn't take an augmented mind to see how advantageous this would be in the job market or the classroom. Super-intelligence would be a total game changer.

Who wouldn't want to be smarter? Who doesn't want their computers to operate faster and more efficiently across a wide range of tasks? The

11. As I have noted before, this is a somewhat complex state of affairs. Proponents of super-intelligence often make it sound like a really powerful computer can think like a person. As David Bentley Hart and others have noted, thinking is a feature that emerges only from conscious beings and, therefore, it is impossible for a machine. The appearance of "personhood" or "consciousness" is merely the product of complex coding that deceives the observer. This begs the question, if the computer appears to be thinking and it appear to be conscious, what's the difference whether it is or it isn't? In other words, people will believe that a complex bit of software actually has consciousness if they cannot tell the difference between a computer's actions and another person's.

appeal of improved intelligence is obvious. But could there be less obvious consequences involved in such lofty pursuits?

Super-Well-Being

What good is being really, really smart for a really, really long time if you are not happy? The third feature of transhumanist thought attempts to create radically happier, healthier people through the use of technology. Let's begin by taking a look at well-being from a bird's-eye view. How have we as a species tried to manufacture more happiness?

- *Religion.* Most people do not think of religion as fun, *per se*, but happiness is more closely aligned with contentedness than fun. Religion places a person into a greater narrative of meaning and belonging, and it tries to make sense of difficult topics like suffering and evil, making them more tolerable.
- *Fostering meaningful relationships.* Since we are social creatures, it makes complete sense that much of our well-being resides in our relations with one another. The company we keep is often chosen for its ability to share experiences with us (thereby increasing our sense of belonging) or its ability to make us laugh (directly increasing our happiness). Beyond the obvious social good of procreation, sexual relationships contribute to intimacy and enjoyment; it is not front-page news to say that sex often makes us more content.
- *Exercise, art, music.* If a person is reasonably satisfied with their life, they may resort to those experiences which give them different forms of extended pleasure. Recreation, creation, and adventure are longtime human pursuits. The enjoyment that we get from art or athletics boosts our mood and releases healthy hormones. Similarly, music stimulates the brain.
- *Drugs.* Ten of millions of Americans rely on the pharmaceutical industry to support their happiness. Prozac is a household brand name. In addition, recreational drugs serve as an escape from the affairs of everyday life, all toward the purpose of making users a little bit happier, at least in the short term.
- *Therapy.* When recreational drugs are not the preferred option, many turn to therapeutic options. Counseling sessions serve as a way to

uncover the sources of discontent, or in a positive sense, to locate some form of inner strength.

- *Eliminate the causes of suffering.* Perhaps the simplest way to increase our happiness is to limit or remove as much physical pain, anxiety, or relational stress as possible.

None of these things, save recreational drugs, are particularly controversial. In fact, these pursuits fill many people's lives with a sense of joy and meaning. But what does it look like to take these pieces and move them into *super*-well-being?

One rather extreme approach is offered by philosopher David Pearce, who argues that humans have a moral obligation to increase happiness and eliminate *all* forms of suffering—not just for humans, but for animals as well. He calls himself an "abolitionist" precisely because he wants to abolish all forms of pain.[12] For this to happen, there has to be some serious reworking of human biochemistry. Pearce thinks a world without suffering is possible through a two-step process, starting with drug therapies. The goal isn't stacking a bunch of short-term "highs" in a row akin to the pleasure someone gets from drugs or alcohol. Rather, the therapy is a laborious process that slowly builds feelings of bliss and suppresses anxiety until the proper biological restructuring has been attained. The goal is nothing short of rewiring our brain circuitry. Once that has been achieved, technology can take us to phase two: gene therapies. Gene therapy, in a grossly simplistic explanation, is the manipulation of genes to treat a disease. If a gene is malfunctioning, the technology allows for its repair or replacement. A complicated and remarkable innovation to be sure, as gene therapies take the place of surgery and drug options. The result? Well, imagine a life free from anxiety—free from the feelings of inadequacy, disappointment, depression and the quick fixes that are required to mitigate these states. It makes you wonder how long the line is going to be at the local gene therapy clinic![13]

Let's expand on this for a minute. We don't just crave happy, healthy, meaningful lives for ourselves, but also for our children. Super-well-being has an intergenerational quality to it. If technology allows humanity to modify genes *in utero*, parents will have the ability to select certain features

12. David Pearce, "The Hedonistic Imperative."

13. A by-product of such an attitude toward health is the way we understand suffering itself. Suffering, in this case, is treated as a pathology rather than as a natural condition that all humans must encounter, endure, and then resolve.

they want for their children. This is not simply a selection of hair or eye color, though that seems to be a popular whipping boy for critics these days. Selecting superficial physical features is exactly that: superficial. But what if a parent could select (for their unborn child) a predisposition for lower levels of anxiety? Or, a genetic modification that makes multiple sclerosis a near impossibility? Or, an enhanced capacity for memory or better reaction time? Parents will face these decisions in the near future. To say "no" to these enhancements in the future might look, to some, like a form of child abuse or gross negligence. After all, no one wants their son or daughter to be *more* exposed to biological or physiological dangers if the parent could all but eliminate them before birth. As if being a parent wasn't stressful enough.

The super-well-being strand of Transhumanism causes us to consider: What does it mean to be happy and healthy? Both have to be taken together for any person to want them. To be happy and not healthy would be akin to drug addiction, where the high leaves you in an ecstatic state while your body falls apart. Likewise, to be healthy and not happy is no recipe for human flourishing either, since most of us deeply desire to find ways to make our healthy life worth living. Eating Brussels sprouts for dinner every night is probably better for you than a big ribeye steak and a bowl of ice cream, but then again, no one has ever celebrated a birthday or a promotion with a heaping pile of Brussels sprouts.

The question is whether or not some hardship is necessary in order to experience a fulfilling life. Protecting a child against all forms of physical and emotional stress, for example, doesn't help the child develop the necessary antibodies (real and metaphorical) to function in the world. There may be other questions you are asking, "Just what qualifies as suffering?" "Does the pursuit of bliss blind us to other worthwhile goods such as the character that develops from delayed gratification?" Important questions to be sure, especially for Christians in light of the cross, since it is through Jesus' death and resurrection that the Christian is made right before God. Spend some time reflecting on the role of pain and suffering in your life, and we will return to the topic in later chapters.

Living in the Clouds

Before I offer a critique of the above Supers, let's distinguish the different strains of transhumanist thought. Not all transhumanists are equal, just like not all Democrats are equal. A spectrum certainly exists, and a helpful demarcation line would divide camps of transhumanists this way: one group generally prefers to *modify* the body and another group wants to *jettison the body altogether* for a virtual or non-organic existence.

The few thinkers I have referenced above fall loosely into the first camp; they tend to promote improvements for the capacities that we already have. They are trying to modify or upgrade the body. For example, Aubrey de Grey seeks longer life—not a fundamentally different form of living. Ray Kurzweil, by contrast, is considerably more extreme. His goal is more ambitious: to predict and embrace new forms of living altogether that allow—even encourage—people to shed their biological bodies for better "housing." The range of beliefs within Transhumanism makes it particularly frustrating to nail down; they themselves recognize how, at times, the movement can act like a moving target. At the end of the day, however, each one of them is committed to improving the human condition through applied technology and reason. And each one of them considers human nature a malleable thing.

Ray Kurzweil is currently the head of engineering at Google. His résumé is a mile deep; he is a recognized leader in digital technologies, futurist thinking, education, and nutrition. Kurzweil has predicted that, by the year 2050, technological progress will have advanced at such a spectacular rate, it's almost impossible to predict what society will look like. In order to back up these claims, Kurzweil applies Moore's Law, a rule-of-thumb that was originally used to predict microprocessor speeds. Today, Moore's Law is used to predict the pace of technological progress; it suggests that technology will double every two years or so.[14] The "law" has been an accurate forecaster of technological progress for the past fifty years, so Kurzweil's faith in its predictive value is well founded. If indeed we continue to see exponential growth in technological advancements, at some point the progress will be so shockingly rapid that the world will fundamentally change as a result. To get a taste of this, consider that, in just six years' time, humanity's entire collection of knowledge will have increased by 800 percent. Can you imagine *that* future?

Okay, an utterly unpredictable future can be a little unsettling, but hardly reason to throw Kurzweil in the extremist camp. It gets better.

14. It appears now that Moore's Law itself is exponential in nature. Technology is doubling at speeds closer to every eighteen months, not two years. In this light, Kurzweil's predictions for future change may actually be a bit on the conservative side.

Kurzweil believes (and he is not alone) that ultimately, digital technologies will allow people to transfer their consciousness onto a computer or digital cloud, allowing their minds to live forever. The physical body of the person is discarded for a machine, allowing the person's thinking self to be passed on to ever more efficient machines into the future. As a robot? Possibly. Or he could just be a totally disembodied being—like intelligence that floats in the cloud. Kurzweil himself hopes for this type of future. He takes a stunning amount of daily supplements in order to maintain his physical health long enough (he was born in 1948) to get uploaded.

For a moment, disregard the questionable plausibility of a mind upload, although Kurzweil is generally respected as a scientist and clear-headed futurist. We will see in the next chapter that online social networks are important because they normalize the break between a person's body and their participation in community—that the body itself was an accidental characteristic of their humanity, not absolutely essential. Now, Kurzweil and others have taken this position to its extreme; the body's *only* function is to house the mind. If one can find another body of tougher, more durable material than human flesh, then so be it. It makes logical sense to escape the most breakable thing we possess—our bodies—for an existence that needs no batteries, no food, no shelter, and . . . no soul? Trans-, indeed!

Secular Faith?

Let's briefly visit an alternate perspective, one that tries to marry life extension principles with Christian view on eternal life. Couldn't the three Supers be accepted as a different but still attainable form of heaven? Shouldn't we, as believers in a new heaven and new earth, be considered transhumanist already? In one sense, yes! There is a lot similarity between Christian and transhumanist ends. For instance, the transhumanist project seeks to extend life indefinitely, provide unlimited knowledge, and give its adherents total bliss. The Christian narrative, likewise, teaches about eternal life (super-longevity) in which all of God's designs will be made known (super-intelligence) in a state where fear or sadness no longer exist (super-well-being). Christian eschatology (doctrine about the end times), however, is promised and made certain by God's decrees in his Word. This is important. Transhumanism can sound an awful lot like a religion complete with its own doctrines, creeds, and view of eternity. But in the absence of a real, external God, the only way to realize such lofty goals is to turn inward and rest on man's ingenuity. And if the past is any indication of how man uses ingenuity to expand his power, there is ample reason to be concerned.

The biblical narrative rejects any utopian future that fails to address the true problem: sin. If transhumanists believe that the primary human problem is its own bodily limitations, then the answers they use to resolve that problem will continue to disappoint. Only a strong view of sin can push our search for a life of grace and contentedness in the right direction. This is why gene therapies designed to increase bliss and happiness are a deficient answer to the question of pain and suffering. They are short-sighted. If the individual is to be understood as a machine, then all solutions are geared toward fixing the inner workings of the person. Just adjust a lever here, another there. In this view, the community is lost. Yet an honest appraisal of humanity's brokenness must include a broader view of the person that accounts for the way we live as families, neighbors, and citizens. Sin is the core problem—both for the individual and for the community. In both cases, Christ must be the answer.

It's only fair to remember that many transhumanists are not blind to the major concerns put forth by its critics, even if those critics are Christian theologians. They readily recognize that technology can be used both to great benefit and terrific evil. However, the internal self-critique may stop too quickly. The Christian's concern is not simply that techs tempt our innate capacities to sin (though they obviously can). It's that the philosophy of Transhumanism automatically creates an unescapable conflict between believers and critics *that is fundamentally theological in nature.* By seeking super-longevity, the transhumanist is simply repurposing the theological language of eternal life toward a secular philosophy. In other words, eternal life is the goal of secular transhumanists and Christian theologians alike. The difference is that Christian eternal life does not come at the expense of your neighbor's decision; it is entirely bound up within the God-person dynamic. This is what makes martyrdom such a profound witness. Though the body is slain by an enemy, a Christian's eternal destiny is never put at genuine risk, as the testimony of martyrs continues to be a source of hope and strength for persecuted Christians worldwide.

For the transhumanist, however, such protections are not available. Transhumanists lack the necessary distance from their philosophy precisely because any real critique of their movement threatens *their* eternal life. If a Christian or Jew objects to radical life-extending advances on religious grounds, the stakes are incredibly high for the transhumanist. If this is the only hope you have for salvation, then no critique will be sufficient for you to abandon it. If Kurzweil survives another twenty years, perhaps he can postpone his death indefinitely. To object to the technologies that Kurzweil hopes will cure his existential crisis is to prevent his secular form of salvation, something he would not be willing to relinquish at any cost. If someone threatened *your* eternal salvation, how far *would you go* to have that threat neutralized?

Discussion Questions

1) What would a world without suffering or death actually look like, this side of heaven?
2) If you had the opportunity to remove certain predispositions from your children, would you? Which predispositions would you choose?
3) Can you think of any portion of Scripture that speaks to the benefits or dangers of technological innovation?
4) Can you identify objections to the transhumanist movement that are *not* religious in nature?

Chapter 2

The Rise of the Social Network

Don't Walk and Text

In 2013, a tourist was strolling down the beautiful St. Kilda's pier in Melbourne, Australia, enjoying the evening air and the pleasant sounds of gently rocking boats. The woman decided to check her smartphone as she walked, slowly absorbing herself in her Facebook feed. Sure enough, she was so engrossed in the tiny screen that she accidentally stepped off the pier and plunged into the water. While unable to swim, she fortunately knew how to float on her back until the local authorities arrived, fishing her out of the water to her great embarrassment. Yes, this is true. Yes, similar events to this have happened all over the world. And yes, it gets worse. When the woman was finally rescued, the authorities noticed that she was clutching the now-destroyed-by-water smartphone for dear life. Even the prospect of drowning wasn't enough for this woman to release the very thing that nearly killed her.

Raise your hand if you've bumped into people while texting. How many of you have snuck in a text while driving, almost swerving into the truck right beside you? I'm guilty as sin. More than that, I'm to the point where every time I leave the house, I instinctually pat my pockets to locate my phone, my mind clutching the device every bit as much as that tourist on the pier. We grab and hold on to the very thing that could lead us into ruin because we tell ourselves that it (ironically) protects us—protects us from isolation, harm, and boredom. But are we just inviting the fox into the henhouse?

By now, we're all familiar with the internet. We know what it does (kind of), and we know how to navigate through it (mostly). If I want a new mystery novel, I forgo the car trip to Barnes and Noble and instead take a short walk to the house computer, type in "amazon.com," and in a jiffy, my new book will be located, packaged, and shipped. I suspect that most of you by now take care of a large share of your family's shopping needs in the comfort of your own home. Combined with the social media craze, this new reality has brought to light an interesting tension that Americans experience every day, even if they are not aware that it exists. On one hand, *homo sapiens* are social creatures; they require relationships and communities forged with strong ties. On the other hand, humans seek efficiency and generally value self-reliance. They try to minimize the amount of time devoted to menial tasks, like driving to the store or browsing the aisles of Target. If we have the option to streamline the process from home, we will. However, by eliminating these trivial affairs we have actually trivialized something more profound: the face-to-face relationships that communities require to thrive. Online social networks have connected us, no doubt, but they have also fostered our desire for instant gratification at the expense of our neighbor. Are we mindlessly bumping into people in more ways than one?

Surrounded by Strangers

Forty years ago, it was not uncommon to know your banker, insurance guy, and supermarket manager by their first names. They made up the peripheral relationships in your life that you wouldn't characterize as absolutely essential but were appreciated nonetheless. These secondary associations increased the overall sense of trust in a community; they kept the town from becoming a group of disconnected strangers. Even though the bonds were not as strong as the ones you felt for your family and your inner circle of friends, simple personal associations such as these helped create social capital.[1] Sure, economies are different now. We do not need a bank because we have a safe form of online banking. We do not need to know this-or-that store owner because we can order goods directly to our homes. But

1. For more on this decades-long trend, refer to Robert Putnam's classic, *Bowling Alone*. He is painstaking about his data collection and his analysis is spot-on.

it's undeniable that such relationships and the social bonds that come with them are quickly disappearing.

It's not just the fringe relationships that are in decline. In past generations, men and women would likely be members of all sorts of different social groups—bowling leagues, quilting groups, bridge clubs, Elks lodges, softball leagues, and dance classes that would include a wide swath of the community. Statistically, these "third places" (social gatherings away from home or work) have been dramatically declining for several decades, and nothing has taken their place as a way to foster local communities rich in interpersonal connections.

The Digital Age is not wholly responsible for the death of more traditional social interactions, but it is certainly speeding along their extinction. Perhaps the question is whether or not local bonds are even necessary in the global online economy. The internet has given us the option to disengage from our neighbors because, for the first time, we can get all of our needs and comforts without leaving our home. With Netflix, you no longer have to go to the movies—you binge-watch *Downton Abbey* and *The Walking Dead* to your heart's content. Shoot, a movie ticket these days is more expensive than a month of Netflix or Hulu anyways! You can watch your show while waiting for your groceries to be delivered, thanks to the kind folks at AmazonFresh. Looking down at your tablet, you transfer money in and out of your Ameritrade accounts while simultaneously scrolling through today's music options. On top of all that, you manage your friendships via iPhone text messages and upload your vacation pics to Facebook or Instagram so that everybody can give you what you want: a big thumbs up. Boom! The completely self-contained life. The *efficient* life.

We can be social and anti-social at once. Pause for just a moment here and consider two simple questions that often provoke complex answers:

In what ways does the digital life *enhance* my personal relationships?

Do online social networks make me more or less social?

I have no idea how you answered, but I do want you to start the habit of actively questioning the value of things you take for granted. This is a start. If we develop discerning habits for relatively simple things like text messaging, we begin to see patterns, connections, and consequences across all of the ways we use technology to make our lives a little bit better.

The purpose of this chapter is to establish a link between the popular technologies of today and the transhumanist philosophies that point our society toward tomorrow. Let's begin by defining a phenomenon that might as well be *the* defining innovation of the twenty-first century: the online social network. Simply put, an *online social network* (OSN) is a web-based gathering of people who develop social bonds by participating in shared interests. The current big dogs are Facebook, YouTube, and Instagram, but endless other platforms allow people to gather, socialize, and give attention to the things that are important them. OSNs operate as bridge technologies. They quietly guide us to believe small truths about ourselves so that it becomes easier to accept more controversial truths down the line. Such formation requires a watchful eye, particularly for the Christian man or woman who is trying to decide whether or not the social networking phenomenon is something that the Apostle Paul counts as "good, noble, or right" (Phil 4:8). With social media, the verdict is a mixed bag: remarkable goods paired with some alarming side effects, like falling off piers into cold water.

The Benefits of Being Connected

In October 2017, *The New York Times* and *The New Yorker* magazine reported that Hollywood movie mogul, Harvey Weinstein, was accused of sexual misconduct by a host of actresses, including Rose McGowan and Ashley Judd. In just a few short weeks, the once-powerful businessman had his reputation reduced to ashes. He was fired by his own company, expelled from the Academy of Motion Picture Arts and Sciences, and faced criminal investigations for at least six different instances of alleged abuse. As the revelations were leaking out, other emboldened women (and men) went public with stories of their own mistreatment, implicating dozens of men (and women) across the vast networks of American entertainment, business, and politics. Online social networks played a powerful role in this public outing. Not only did the initial accusations emerge from social media platforms, but the legs of the story were sustained by hundreds of thousands of subsequent retweets and Facebook posts. No one could hide from the story. Everyone wondered, "Who would be exposed next?"

Perhaps the most fascinating by-product of the scandal was the #metoo campaign. It began when actress Alyssa Milano encouraged the use of this hashtag as a way to bring to light the widespread nature of misogyny. She essentially posed the question, "Have you been a victim of sexual misconduct? #metoo." Millions of women who had been victims of similar behavior joined in solidarity by posting #metoo on their Facebook status updates and Twitter feeds.

The effect was so staggering that much of Hollywood went underground. Male executives, actors, and agents have been running for cover like fish from an approaching shark. OSNs, in this case, levelled the playing field. Just fifty years ago, three networks dominated the entire distribution of news. Editorial decisions, including *what actually counted as news*, were made by a handful of power brokers—almost exclusively men. The New Media has obliterated this monopoly. OSN users have become the new curators of content. Now, anyone with a camera phone (i.e., everybody) and a penchant for hot takes can become the next big thing, or more commonly, *point us* to the next big thing. If that "hot item" is social awareness about a controversial issue, then so be it. Anybody can open up a news story.

A balanced approach serves us well here. If you are like me, it's easy to slide into "the sky is falling" pessimism about the state of our world. I admit that I often gripe about the newest online technology or app, complaining that it is doing this or that to our humanity, with little or no actual support, and actively ignoring the tremendous value they bring to our lives. The above story about the #metoo movement is but one example of using an online social network toward a potentially positive end. A healthy tactic is to recognize when OSNs (and things like Google) are actively contributing to our society then apply some critical thinking to examine their weaker points. There is no need to become an anti-tech vigilante. At least not yet. Instead, let's begin by recognizing the goods of the major players, and then look for the possible consequences that come with uncritical use.

Social media levels the playing field in so many ways. YouTube allows for the aspiring filmmaker or comedian to test material at a grass-roots level and to generate some positive free publicity. Instagram provides similar opportunities for budding social influencers, photographers, and graphic art designers. Video conferencing platforms like Facetime and Zoom have allowed millions to partake in virtual conversations that include (to a limited degree) the non-verbal communication that makes face-to-face conversations so personal.[2]

Generally speaking, the benefits of OSNs can be boiled down to two fundamental goods: connectivity and information. All the other benefits

2. The popularity (and necessity) of video networking has never been more apparent than in the COVID-19 outbreak. During the nationwide lockdown, many turned to these applications to keep connected with their family and friends.

are derived from these two. We desire to be as connected as possible with access to as much information as possible. You may be thinking of some counter-points here, but for the moment, place yourself on this theoretical sliding scale:

> On a scale of one to ten, would you rather be disconnected (1) or connected (10)?
>
> On a scale of one to ten, would you rather have no information (1), or all the information (10)?

Nobody wants to be ignorant. Nobody wants to be utterly and completely cut off from their community. But that's not the whole story, either, is it? For Christians, the desire for connectivity and the promise of information can quickly become an *ultimate* good, as if our iPhones fulfill the obligations and responsibilities of an all-knowing god.

The Digital You

Several weeks ago, my family and I frequented one of our favorite little holes-in-the-wall to enjoy a heaping plate of shrimp tacos. As I glanced over my wife's shoulder to the table immediately behind her, I noticed what felt like, in this day and age, a "strangely normal" sight. Four college-aged women were seated together in complete silence. Their food had arrived, but instead of eating—instead of talking or laughing—they all had their heads buried in their laps. Their smartphones required their immediate attention. Not surprisingly, they all had vague smiles on their faces, amused by one of a million different entertainments that smartphones make immediately available.

I suppose I could interpret this scene in one of several ways: (1) perhaps they were enjoying a YouTube clip, game, or website together with the other members of that table, making the experience a shared, community-building exercise. Or (2) each person may have been engaged in a text conversation with someone not immediately in her physical proximity, thereby giving one group (the table of friends) her physical self and another (whomever the she was communicating with privately) her intellectual or social self. Even more puzzling, they may have (3) been texting each other as a semi-discreet way to have side conversations with other members of the table. No matter which option is taking place, something formative is happening. Their eyes and their attentions are being diverted away from the embodied (i.e., the physical) to the disembodied (i.e., the digital or virtual).

I call this episode strangely normal because I suspect you have had this exact experience or something close to it. The scene is becoming an all-too-common one. These instances of "split attention" are curious for a variety of reasons. A set of hidden assumptions are inscribed in social technologies that make this behavior odd, or better yet, *at odds* with some of our deepest held beliefs about human behavior. OSNs rest on the assumption that you can participate in multiple forms of social life at once without fragmenting yourself in the process. One can, quite literally, be in two places at the same time, albeit in different forms.

Virtual reality (VR), even more than surfing the web, is an example *par excellence* of the simultaneous interplay of various forms of existence. VR blurs the distinction between that which is "real" and that which is "virtual"—a distinction that is quickly failing. Today's youth do not, by and large, separate these two types of existence; whether moving through physical or digital environments, they have enough presence in each to be comfortable. In a few years' time, everyone will be including VR rooms in their home design as a natural extension of this comfort, spending hours in virtual worlds, traveling, gaming, and socializing without ever leaving the house.

And here's the rub. On one hand, we give lip service to the idea of being in multiple places at once. We might, for instance, say during a friend's time of testing or mourning, "I'm with you in spirit." Yet what we generally mean by this statement is that we include this person or that situation in our prayer consideration, not as actual presence, but as a topic to think deeply about. Our *considered thought* about the subject sufficiently acts as a substitute for us not being there. On the other hand, OSNs (including VR and video-conferencing) are actually placing us in positions to be connected in real-time conversation with several people in several different locations at once. This leaves us in a serious tension: Can I divide myself into parts? Can I give two social groups my simultaneous attention without losing the important pieces of what makes me, me?

We have taken multi-tasking to its extreme. A young mother of three is essentially a juggler, trying to keep all the balls of her life in the air at the same time. She tends to the near-constant needs of the children and keeps the family's finances in order, along with a hundred other tasks that must be done often while holding down a job of her own. Multi-tasking is the name of the game if she wants to survive. However, what we usually call multi-tasking is not the executing of simultaneous tasks *per se*, but rather, the ability to complete many tasks stacked up so closely that it appears like you are performing the work of a dozen people.[3] I bring this up because at first glance the constant maintenance of friendship bonds in social media looks like a complicated form of multi-tasking and therefore has relatively few negative consequences. The modern mom, for example, is simply being efficient with her limited resources.

But in the case of social media, I'm not convinced the users are really multi-tasking. Return to the college kids at the restaurant table. The communication that is happening in this instance *is simultaneous*: they are physically communicating with those they are present with in person, and yet they are also intellectually or emotionally connecting with the person with whom they are texting. There is no stacking of tasks, if you will. It's not multi-tasking, it's "multi-talking." Online social networks nudge us to embrace this multi-dimensional communication since the oft-stated goal of all OSNs is a more connected, more informed world. A man who fires off a text to his wife while at a business meal isn't trying to be discourteous to the people with whom he is eating; he is still physically present with his dinner guests. He might not even realize the type of attention splitting that is taking place—it's become, ironically, *too natural* for him to realize it.

These are relatively harmless examples. I would argue that most people, when asked, would deny that it's possible to split yourself into unlimited little bits. The way we use online social networks, however, seems to cut against that belief. Open up your smartphone. Which applications tend to split *you* into little bits? Have you noticed their influence, or do you simply accept it as the cost of being a well-connected man or woman?

OSNs foster another subtle assumption. We have an ever-present audience. For those who partake in the weekend fun of karaoke, this is not a foreign concept. The audience—the energy of the crowd—gives life to the performer, whether the performance is a rousing rendition of Celine Dion in the karaoke bar or a carefully timed status update on Facebook. Everyone wants the likes. And to get likes, you must be followed. The desire to be followed (and liked) is particularly strong in teenagers, who have just enough

3. Many scientists are calling into the question the mere possibility of multi-tasking. The brain is not biologically wired to process dual information simultaneously. This would, in part, explain the danger of driving and internet surfing at the same time or why the sleight-of-hand techniques of a magician are particularly difficult to solve. See Detweiler, *iGods*, 62–68.

narcissism to think that their every move is being watched and analyzed by their peers. Online social networks are exacerbating this effect. "What *is* new is a technology that takes our naturally adolescent assumption that the world is watching, and offers us a spotlight, a microphone, and a stage as vast as cyberspace from which to act out our assumption—with our legion of friends serving as an invisible entourage."[4] Adults are just as guilty. It does not matter what age you are; everyone wants someone to pay attention to them. Here lies another tension: we only act like we have an audience when we are engaged in the digital world. You don't walk around town (unless your last name is Kardashian) thinking everybody is enthralled with your choice of leggings, just waiting for you to say something worthy to be put into print. But then again, how many of us post pictures of our latest outfit? Our last crazy good meal? The digital world's gaze seems to be growing with every passing day; we allow it to penetrate areas of our lives that were formerly private, off limits to all except maybe our inner circle. Now, all bets are off.

This new reality might just reveal a significant truth about the giants of Facebook, Twitter, and Instagram: they are popular, in part, because they allow us a slick avenue to plea frantically, "Pay attention to me!" Whether people actually follow us online or not, the fiction we tell ourselves is a strong one: in order for people to like the real me, I better do some editing.

Hidden Goods

Imagine with me, for a moment, a scene from a bygone era. It's early autumn and it's finally time to gear up for the cold slog of winter. One morning, you notice that your woodpile is less than sufficient for the upcoming months and resolve to do something about it. You call up your buddy, Carl, and ask him to join you for a day in the woods collecting and chopping some pine. Picking him up in your truck, he hands you a thermos of coffee and asks if you watched the game last night. Half an hour later, you have arrived at a section of forest with some fallen trees. You chainsaw through the seasoned wood, covering both you and Carl in sawdust, but you don't mind because your work is shared with a good friend.

With the truck bed full of timber, you head back to your home. Splitting the wood comes next, along with a cold beer and details from Carl's deer hunt last week. By the time the last log is split, you have invited Carl's

4. Rice, *Church of Facebook*, 111.

family over for dinner that night as a gesture of thanks, and you don't mind at all because it gives you an excuse to smoke some brisket. As you walk past the newly stocked woodpile, you find yourself strangely happy even though you are absolutely filthy.

Now reflect for a moment:

What are the goods here?

To what degree is something's worth a function of the time it takes to get it?

What is more important, the end or the means?

This digital world is a true rival for our collective attention because it offers us a variety of goods often unavailable to the "real" world with remarkable speed. We have instant communication, instant information, and instant gratification. Just one look at the simple act of texting reinforces my point. When we receive a text from a friend, we get to (1) hear from someone we care about, (2) get information about how they are doing, and (3) experience the splash of dopamine that courses through our brains when our phone buzzes with a new message.[5] Personally, I have contests with my friends to determine who is best informed with the latest sports scoop. I like to get the news first and text it to my friends as quickly as possible. "Did you hear that ______ got suspended for three games?" "Looks like ______ broke his ankle. He's out for four to six weeks!" In the world of the twenty-four-hour news cycle, my friends and I are active and willing accomplices. The final product (information) is the only value, never the work to acquire it.

All Games Aside

OSNs such as Facebook, YouTube, and Instagram are ground zero for these "instants," but video gaming (a slightly different form of OSN) is emerging as a powerful gathering place to receive such goods as well. There is a growing body of research that suggests online gaming provides more social goods than just base entertainment. Gamers are attracted to the vast narratives that many games provide as well as the built-in challenges that provoke deep thinking and creativity. Some researchers go so far as to suggest that games, in particular online or video games, might offer a framework for solving some of society's greatest challenges because of the nature

5. For a fascinating and accessible look at the physiology behind OSN use, I highly recommend Nicholas Carr's work, *The Shallows: What the Internet Is Doing to Our Brains.*

of gaming itself. To make any game enjoyable, it must involve a series of obstacles, require creative thought, and push the player to use relational resources to achieve "an epic win." How many people can say that their job offers this collection of prized outcomes? Now you know why 180 million Americans retreat to their games as a way to escape the doldrums of the real world. Reality just can't compete.[6]

People are retreating from a dull and unsatisfying real world, spending much or most of their time in digital worlds designed to keep their attention. The everyday hardships of jobs, taxes, and bills serve as reminders that the world isn't built to bring out the potential in us. Consider this argument in the context of a short vignette: Imagine you are a thirty-year-old single male working at a white-collar, data entry job for forty hours a week. Would you rather be known as Bob, the data entry guy who has no appreciable status or influence . . . or be known by your online gaming name, (*insert cool alias here*), where you are the overthrower of kingdoms, master of weapons, wielder of ancient spells, and widely known as a top player in this-or-that gaming genre? You might choose the former for a variety of reasons, but perhaps you can see why the latter is so appealing. By simple odds, about three out of every five people in your workplace spend regular time playing online games.[7] Did you just have a cup of coffee with a level sixty wizard without knowing it?

There is no doubt that a person's retreat from their embodied neighborhood has huge repercussions on their sense of identity, an identity that has been largely shaped by the major OSNs and online gaming industry. Maybe we've been thinking about this wrong all along. What if we turned around the question: "If millions of Americans are using games as a way to enhance their life, what can we do *to make reality more like games*?"

Footing the Bill

Play the gambler for a moment. If you were to initiate a conversation over lunch, what would be the over/under in minutes before you were interrupted by one or both of your cell phones? Four minutes? Less? Even if the

6. Jane McGonigal, a game designer, believes gaming could be a societal force for change due to the intrinsic qualities found in every successful game. See McGonigal, *Reality Is Broken*.

7. McGonigal, *Reality Is Broken*, 3. McGonigal estimates that there are 183 million active gamers in the United States who play, on average, thirtten hours a week.

phones don't make a peep, I'd wager a pretty penny that you'd check your phone once or twice in the first few minutes anyways, as a simple matter of habit. Interruptions by our devices are now the norm rather than the exception. 'Tis the cost of being in the loop, yes?

There is a distinctively transactional character to the digital life. For every gain that is made in efficiency or entertainment, something is lost. Or, perhaps better put, something is required. "FOMO" is one of those payments. The Fear of Missing Out. Being perpetually connected affords a user a stunning amount of information at their fingertips, whether it's the news (e.g., there was a bombing in Madrid twenty minutes ago!) or the constant connectivity that is nowadays required for friendship maintenance. The digital person is perpetually *on*. Freedom and rest are in short supply when the iPhone becomes a tether rather than a tool, when it's the last thing you look at before you go to bed and the first thing you check when you wake up in the morning. Teenagers and adults alike are so invested in this FOMO that, when separated from their phones for an extended amount of time, the person experiences measurable increases in anxiety. Believe it or not, this unease consistently peaks on certain days of the week! Cell-phone-free Thursdays, in particular, provoke the most stress as this is the day when most weekend plans are developed. To be separated from the phone is to be separated from the community of friends and their Saturday night adventures. While you are "off," your contacts are theoretically having a conversation in your absence, strengthening their connective ties with one another, and leaving you in the dust. All because you left your phone on the kitchen counter.

But I'm not certain that *separation* is the ultimate root cause of this fear. Ultimately, it's the fact that we are limited beings. We're not gods. We're less comfortable with our limitations now that technologies have caused us to rethink what is possible. Our bodies can only do so much and our time is short. Once you've been given a taste of an unlimited life (at least in a small way), it's difficult to put down that drink. After all, consider how *incomplete* you feel when you put down your cell phone!

The Bridge

Time to put some pieces together. The central thrust of this book is to take a hard look at digital technology and transhumanist principles through a thoughtful Christian perspective. Let's not kid ourselves, all of life is theological. That is to say that everything we do is enabled by the God who both created and sustains us. Nothing falls outside of the gaze of God, including the spiritual effects brought on by OSN use. The consequences of ignoring these influences are serious. While I haven't yet brought heavy theological resources to bear in this discussion, I promise that it will come soon. Right now, I'm just trying to get you to observe the landscape. We cannot survey the architecture of Transhumanism until we first examine the concrete foundation and check for cracks. In this chapter, we have been looking at social technologies to give us an "under-the-hood" view of this foundation.

Social networks are the gateway drug to Transhumanism; they are cheap, easy-to-use, and powerful. Okay, that might be a bit harsh. At the very least, OSNs force us to make decisions about how best to use our social energy. We may decide that the ease of OSN friendship maintenance is worth some of the side effects, a valuing of connectivity and efficiency over more difficult, embodied relationships. Or we could decide that the price is simply too steep and conclude the only way forward is to quit technology cold turkey. If we regard OSN use as *transactional*, we can be more discerning and intentional in the ways social media is used and avoid throwing out the baby with the bath water. Remember that OSNs are offering an alternate reality that is viewed by many as more beneficial, enjoyable, and rewarding than life in the "real world." If only real life had a button that "added friends" or, ironically, simplified your love status to a vague "it's complicated."

So what's the link between a harmless status update on Facebook and the potentially life-altering claims of Transhumanism? First and most importantly, OSNs teach us that it is normal to divorce one's physical self from ordinary human interactions. The body is no longer necessary to participate in everyday relationship-building. Just send a text and it's like you were beside your friend all along, no matter the actual physical distance separating the two of you. Just seeing her Instagram post makes you feel like you were, in some small way, part of her vacation. By doing this, however, one implicitly accepts the idea that the body really doesn't play a significant role when it comes to relationships. If you're not buying this quite yet, no problem. We will spend some more time on this in a later chapter. Technology provides the perfect opportunity to overcome our limitations; that's what technologies *are designed to do*. Since my body cannot be in two places at once, I can divide myself into several parts—physically present here while intellectually present there—all while believing that each interaction, each conversation, and each person I interact with still gets the whole Me. What enormous power.

Second, we do not use OSNs simply to maintain relationships and seek out information. Social media sites are now ground zero for personality manipulation and maintenance. In other words, people begin to use these platforms as ways to construct their identity—a public place to experiment with how they look, what they prefer, what they stand for, whom they associate with. This is an explicit acknowledgment that the online world can help us discover who we are as well as help us be who we want to be. There is nothing fixed about human nature or identity in this model, which means we're not too far from the philosophy of Transhumanism and its comfort with a perpetually evolving humanity.

I am not suggesting that everyone who uses Snapchat is destined to fall in love with the transhumanist vision of the future. What I am saying is that our widespread acceptance of digital media has laid the foundation for a view of human flourishing that is quite different from past generations, and this change is substantial enough to merit our scrutiny. We align our social norms, laws, and customs to reflect what we think makes for a good society. If OSNs are actually changing the way we think about ourselves (i.e., who we are, how we fit in), it is inevitable that our culture will begin to shift toward a future that accommodates such a view.

OSNs and Human Needs

Social media isn't going away, nor do we necessarily want it to. But what next? By taking the gains and popularity of OSNs seriously, I believe individual Christians can honestly evaluate the transactional nature of social technologies to ascertain their benefits on a case-by-case basis. The popularity of the OSN reveals something important about the fundamental needs of the human person. Technologies are most desirable when they fulfill roles that people cannot—when they can assist a person to accomplish a task that would otherwise be difficult, even impossible. We can say with confidence that web-based forms of connectivity and identity-building strike a distinct chord in humanity precisely for this reason. To stay connected in this day and age, complex tools are needed. Online social networks make it easy to find places of belonging in spite of physical restrictions. Acceptance can be found in the relative anonymity of cyberspace. OSNs also make it easy for a user to experiment with their public persona without the pressure of face-to-face encounters. Trying on a denim romper might be a risk in many social settings; in the OSN world, one can try on a new *gender* without anyone ever knowing.

Since nearly all of our OSNs incorporate personal identity-building into their programs or applications, we are just a short distance from a transhumanist-friendly position. Facebook users, for instance, have become accustomed to the idea of branding themselves (i.e., their avatar, their preferences, and their carefully curated photos) in disembodied, online environments. This is not to cast any moral judgment on the practice; I am just pointing out how many people are completely comfortable with the *image* of oneself as the standard form of identity. I spend more time than is necessary when I pick my Facebook profile photo precisely because of this fact. I know that people will look at that photo and use it to determine who I am and what I value. I'm guessing that many of you have gone through similar experiences.

The popularity of these networks, as I mentioned before, allows us to make some significant comments about our ongoing question, "What makes humans, human?" All of us need a measure of community, a place to feel loved and accepted for who we are. This is both a statement about knowing (we want to be known) and about neighborliness (we want to be with others). Even for the introverts among us (myself included), full-bore isolation would eventually drive us crazy. We simply need the physical proximity—even the physical contact—with the people in our social circles if we want to experience the full range and beauty of the human experience. The question is whether our experiences in the near future will actually be human at all.

Discussion Questions

1) Overall, do you think that online social networks have made us more connected, or more isolated? Can you think of specific examples that support your conclusion?
2) What are some of the observable effects (good or bad) of online social networks in your life?
3) What does the advent of the OSN say about humanity other than the things mentioned above?

Chapter 3

Why We Love Shiny New Toys

Beneath the Surface

IF YOU HAVE SPENT any time in a history class, you will know that whenever you study a particular event, there is the easy explanation and the hard one. An easy explanation might go like this.

"How did World War I get started?"

"Well, Archduke Ferdinand of Austria was assassinated in Sarajevo in 1914, setting off a declaration of war between Serbia and Austria-Hungary. All of Europe took sides. Ergo, war."

Is this explanation true? Well, yes, to an extent—but it certainly doesn't tell us anything useful about the complex political and economic conditions of Europe at the time. In fact, the assassination may have only been a very small fuse on a very large bomb. A lit fuse without the bomb is comical. Equally comical is trying to explain a huge crater in the ground by talking about the finer points of fuses.

A more difficult, nuanced approach is necessary. The hard explanation attempts to account for a multitude of variables. The deeper you probe into the world of explanations, the more you realize that tidy answers are often far too simplistic, and they are just as likely to be misleading. It's incredibly difficult to predict, let alone explain, the outcomes of American football games with twenty-two moving pieces on the board. How much more difficult it is to lay out a convincing explanation for geo-political events!

So here is the question we are going to tackle: *Why* do we love our digital technologies? This is a simpler form of a question that the previous chapter might have prompted you to ask. That is, why does Transhumanism have a cultural foothold in the first place?

The easy answer to the first question might include: (1) They are fun, (2) they make life a little easier, and (3) they allow us to communicate with others and stay connected, and so forth. These answers may be true, but they are ignoring the deep history of how people (particularly in the West) have come to accept them as a part of the good life. The hard answer, in part, will be the subject of this chapter. We will be examining the roots of the tree to better understand why this philosophy has captured the attention of Western culture. Just because the task is difficult does not make it impossible; I will attempt to identify some key markers that may not be utterly comprehensive, but at least they will be able to point us in the right direction.

In chapter 1, we tried to get a handle on the central features of Transhumanism, so it would serve us well to consider the fundamental forces that have shaped such grand visions. In one sense, I have hinted at the end of the story, a future filled with super-charged humans who have far greater "powers" than we currently possess. Now it is time to tell the beginning of the tale. Just as technologies are usually built on the backs of other innovations, so often ideas are birthed as improvements of prior ideas. First, I would like us to consider the way ideas are introduced and how they interact with the establishment. I will also toss in some observations that may resonate with your individual experience with technology. Second, we will widen our gaze historically. We will see how the historical victory of certain ideas has created a broader movement, a movement that is so significant that it has the feel of a story. This tale has been called the Myth of Progress, and like all myths, it is a special type of narrative that explains *why* we are *how* we are.

Orthodoxy vs. Innovation

To be human, in part, is to understand that we live amongst other humans. Another way to say this is that we recognize that we interact with other minds—other personalities or perspectives—that may or may not see the world as we do. Adults know that people see the world from differing points of view, so the presence of other minds shouldn't come as a shocker.

Conflict comes about when two opposing ideas about the world meet face to face. This is easily observed in marriage: "Why do I always have to do the dishes?" "Shouldn't we being raising our kids *this way*?" It's easy to create echo chambers in our lives where we surround ourselves with people who think like us; Facebook has made friendship selection an art, after all. Even for the most dedicated Facebook user, however, confrontation simply must take place at some point.

The phenomenon is particularly noticeable in American politics, where one side of the aisle embraces a view (say, about immigration or taxes) that is at fundamental odds with its opponents on the other side. While the disagreement may appear to be about a particular piece of legislation's worth or lack thereof, I would argue that in many instances the conflict comes from incompatible answers to foundational questions. Questions like: What makes for a good society? What makes humans uniquely human? Can morality be legislated? Foundational philosophical disagreements necessarily create discord down the line, at more public levels. Viewed positively, such disagreements can lead to healthy discourse and widen the perspective of everyone involved. Negatively speaking, political disagreements lead to *ad hominem* attacks, mudslinging, and character assassination.

If we stand back just a little bit, I think we can see how a large portion of our understanding of the world is formed by a great tension between the camps of Orthodoxy and Innovation. Let's dig down and look at the roots of the tree where the foundational tension resides.

Orthodoxy is an understanding of reality that is based on the acquired and established wisdom of the past. The term has been used to designate certain forms of religious practice (e.g., Orthodox Christianity or Orthodox Judaism), but the word itself does not have to refer to religious belief at all. Orthodoxy is simply normal, accepted practice. Alternatively, you can think of orthodoxy as a synonym to tradition or traditional ways of thinking.[1] Orthodoxy, like tradition, is built on a trust relationship with prior generations and can include rituals, doctrines, even frameworks for how we think about the world. I make chili the same way my father makes it. I have simply trusted that his way was *the best way* to make chili (no beans, *ever*). Since my experience verifies it as delicious every time I take a

1. In the ways I am using the terms, tradition is more about practice and orthodoxy is more about the way we think about something.

bite, my trust is confirmed and no further recipes are up for consideration. Orthodoxy, at its core, is built on trust. Trust in the wisdom of the past.

You may already be imagining a man in your office or a young lady on your committee who takes seriously the contributions and insights of the "old guard," believing its long-established take on reality is a worthwhile one to continue. Perhaps *you* are that person! I admit that orthodox views are generally quite attractive to me. Perhaps it is because of the life experiences I've had or the interactions with tradition (and traditional ways of thinking) that have been, on the whole, positive.

For every yin, there is a yang. Even the most traditional position on anything was initially birthed as an idea. Ideas are the creative driving force behind change in our world. They challenge the *status quo*. They excite us by opening up new possibilities. They also can be deeply unsettling. Victor Hugo once wrote, "One resists the invasion of armies; one does not resist the invasion of ideas."[2] As a professor, I'm a trader in ideas. I make a regular habit to tell my students that good universities are not safe spaces by definition precisely because they are in the business of discussing and evaluating ideas—and ideas are fundamentally dangerous. All ideas are guilty of aggravated assault; they seek to overturn, to delegitimize, even destroy the old way of doing things. Video really did kill the radio star. The process of *innovation* (or the implementation of new ideas) seeks fresh answers and insights into the nature of reality, putting a great deal of pressure on traditional ways of knowing.

A centerpiece to innovative thought is the assumption (a safe one, I think) that the world is constantly changing. Even though answers from generations past had some usefulness, the world today is simply different. New systems must emerge based on the simple fact that societies develop, grow, and evolve. Orthodox thinking, by contrast, is an effort to rally together an understanding of the world that withstands the ebbs and flows of cultural change. Orthodoxy and Innovation, then, exist in a near constant tension.

The Swinging Pendulum

Countless examples of this tension can be found in history, but perhaps the most famous came about in the early seventeenth century, when Italian polymath Galileo Galilei provided a defense of Copernicus's heliocentric

2. Hugo, *History of a Crime*, 627–28.

program over and against classical geocentric models supported by the Roman Catholic Church. The idea of a static (not moving) Earth was seen as a threat on multiple theological and philosophical fronts, calling into question how a scientist should approach the pursuit of truth beneath the twin behemoths of religion and science. For the church, the authority of Scripture was at stake, as well as man's orientation within the created order.[3] Galileo, a Christian, understood the implications of Copernicus's work, as well as his own observations; to affirm the Earth's constant movement in space would upset a Ptolemaic system that had been "orthodox" for hundreds of years. His trial and sentencing serve as an example *par excellence* of the tension to which I refer.

Fast-forward to present day. Transhumanist philosophies can be just as unsettling. This is not to say that they will be ultimately vindicated and widely embraced, but rather, they threaten to undermine the specific place and role of humanity that the Bible offers. Christians, conversely, need to be prepared to answer: (1) if this threat is imagined, where can a Christian lend support to innovation and avoid the charge of Luddism[4], or (2) if the threat is real, what to do about it without losing their Christian charity. Ultimately, I am asking whether or not Transhumanism should be treated as an innovative philosophy that can benefit the church before the church takes to excommunicating the ideas just for being ideas, lest Christianity falls into the same trap it did when it reacted poorly to Galileo.

I would argue that the tension between orthodoxy and innovation is a necessary and fruitful one. Each side keeps the other from swinging too far, as there is a danger latent in each. What do I mean by that? In the case of orthodoxy or tradition, if someone or some institution takes this position to its extreme, there is the danger of *legalism*. Legalism is the creation of laws where none need exist. Legalistic people are the ones who move from "I prefer things such-and-such a way," to "This is the only way things can be done." In the Bible, the Pharisees pursued a form of legalism. In their

3. The common mistake is to think that the Ptolemaic system sought to protect humanity's "lofty" status over and against the rest of the created order. A better way to state the theological concern here was to say that humanity was the focus of God's redemptive efforts, and Christ was sent explicitly as a human to save humans. This better reflects the theological and philosophical reasons for protecting a geocentric model.

4. Luddism originated when English textile merchants in the nineteenth century violently opposed the introduction of the automated loom. Seeing their profession in peril, the rebels destroyed property (including the looms) across the countryside. For this reason, a Luddite is a person who rejects technology. In modern language, a Luddite (or neo-Luddite) is a person who rejects a lifestyle inundated with digital technology.

encounters with Jesus, it became obvious that the Pharisees loved the Law (Torah) more than they loved God. Those who favor orthodox positions can quickly descend into a form of Luddism—that is, they can react violently against any form of new thought or innovation simply because *it is new*. In terms of our overall discussion on technology and Transhumanism, I think this approach is wrong-headed and far too extreme. A faithful Christian can maintain a healthy respect—even a preference—for traditional cultural views without descending into legalism.

Innovation itself is not without danger. If a person or cultural movement casts off all connections with the past, the risk is *heresy*. Heresy is an old-fashioned word, no doubt. We don't use it too often for fear of offending people of other faith traditions. I'm using it, however, as a way to describe how an idea can explicitly harm core convictions. Take, for instance, the case of Jesus. Christian orthodoxy had held, for over two centuries, that Jesus was both fully God and fully man. Arius, a fourth-century bishop in Alexandria, began to teach an innovation that Jesus was created by the Father, not fully divine as traditional Christian doctrine had professed. Since this utterly contradicted the divine claims of Jesus in the Gospels, this idea found itself squarely in the camp of heresy.[5] Christianity as we know it could not have survived if Arius's teaching grew to be the new doctrine of the church, precisely because the very authority of Scriptures (where we get our information about Jesus) was directly undermined.

Heresy does not have to be associated with religion—at least not in the way I am using it here. Take for example, a company that manufactures car safety products. The expressed vision of the company president was to provide the very best in life-saving equipment in today's vehicles. Now imagine this company's profits begin to waver, for whatever reason. At the board meeting to discuss options, a young executive in the company suggests retooling the machines to manufacture bullets instead. Surely, this qualifies as an idea. Perhaps, it's even an innovative idea. But because the vision and *ethos* of the company expressly communicates the purpose of life-saving products, the manufacturing of ammunition would be considered heresy.

The word "heresy" tends to scare Christians a little—for good reason. How can a Christian preserve this orthodox-innovation tension without sacrificing the core features of Christianity?

5. The divine claims of Jesus are well-documented, though often require a little hard work to uncover. Perhaps the boldest of the divine claims comes in John 8:58, where Jesus connects himself to the Old Testament revelation of God's name in Exodus by saying, "Before Abraham was, *I am*."

God as Both?

There is a temptation to think God solely resides in the orthodox camp. Perhaps this is because we refer to the basic, standard package of Christian belief as "orthodox." The Apostles Creed, for example, is simply a set of traditionally held statements about the Trinitarian nature of God.[6] Likewise, the Old Testament is filled with precise ways of referring to God and his qualities: he's one, eternal, unchanging, zealous, and on and on. If we go down this road too far, however, we lose sight of the portions of Scripture where God undeniably acts as an innovator. For example, in the Old Testament, Hosea prophesies on behalf of God to the Israelites: "For I desire mercy, not sacrifice, and acknowledgment of God rather than burnt offerings" (Hos 6:6). If we run by this statement too quickly, we lose the absolutely remarkable reversal that happens here. The entire structure of Jewish life was built around a series of sacrifices that assured the Israelites a right relationship with God. God himself set up the system! Read the first few chapters of Leviticus and it becomes clear in a hurry how much God cares about ritual cleanness and proper sacrifice etiquette. His attitude toward his people appears to be governed by the people's proper use of animal and grain sacrifices. Now, Hosea is suggesting that God's heart is primarily concerned with mercy toward the neighbor—not sacrifice! Is it possible that God might be offering a new understanding (innovation) about the relationship between God and person, and by extension, person to person?

In the New Testament, we have another poignant tale of God's surprising movement in the world. In Acts 10, Peter falls into a trance and has a vision. He sees a great blanket let down from the sky, filled with all sorts of animals normally thought of as unclean by Jewish law. A voice tells him, "Get up and eat!" Peter refuses, deferring to the cleanliness standards of the Torah. It's a perfectly devout response to God's invitation. Yet, God says, "Do not call anything impure that God has made clean" (Acts 10:15). How is the orthodoxy-innovation tension playing itself out here? Peter is holding onto to the orthodox understanding of God's will concerning purity, yet God seems to defy his own Law by commanding Peter to eat. Once

6. When I use the term orthodox, I am not referring to the proper noun used to define the Eastern Orthodox branch of Christianity over and against the Latin West. I am using it in its most basic form as a way to communicate traditional, standard, time-tested ways of doing things.

again, God is the innovator. He is the God who unfolds. Creates. Reveals. Surprises. He is the one who "makes all things new" (Rev 21:5).[7]

While I am not suggesting that God changes his mind on a whim, I am opening the possibility that his will is revealed in surprising ways, often creating a healthy tension internal to belief itself! The unfolding of God's will allows an older command to be fulfilled or altered for the present, much like a parent adjusts the house rules for children as they age—sometimes the rules are adjusted, other times they are thrown out as obsolete. Such transformation serves to advance God's design for his people.

Progress and Story

Christianity has a muddled relationship with innovation and progress. On one hand, Christianity has played a significant role in the rise of modern science. From its origin, Christianity has taught that the universe is God's creation and, therefore, it is good and orderly. Any investigation about the world then is an investigation into the very goodness of God's creative work. Another way to say this is that God has left us two ways of knowing him, natural knowledge and revelation. Though certainly flawed, our experience of the natural world does, in fact, reveal certain limited truths about who God is.[8] Part of our Christian mandate as relational beings is to create culture out of the natural world.

On the other hand, the relationship between science and Christianity has been somewhat strained as scientific advancement has often called into question the historical claims of the church. Galieo's case above serves as an example. The church's commitment to man's primacy—based on a particular reading of the Bible—led it to make broad sweeping conclusions about

7. Peter's dream ultimately reveals God's intent to include the Gentiles (formerly unclean pagans) into the plan of salvation. The Apostle Paul calls this inclusion a "great mystery." See Eph 3:4–6.

8. It may be useful to point out a classic distinction between natural knowledge of God versus revealed knowledge. Natural knowledge is any information we can get about God that comes through the observation of nature or the application of our reason. For instance, we may be able to discern God's sense of order by observing the mathematically precise movements of the planets and stars. Revealed knowledge, by contrast, comes to us from the outside. It is what we know about God through Holy Scripture and the person of Jesus Christ. The former is a notoriously weak apologetic for the specific character of God, revealing little (if anything) about his disposition toward humanity or his plan of salvation. The latter is the foundation of the Christian church and is, unquestionably, the strongest foundation for Christian doctrine.

cosmology. Most Christians reject the notion that the Bible serves as an all-encompassing authoritative textbook on astronomy, yet scientific challenges will continue to press the attack on other fronts. Cultivating a sense of discovery certainly has a place in the Christian worldview, but we should deliberate carefully whether or not discovery is a pursuit without limit. Is it okay for humanity to "improve" or augment God's creation? For instance, should elective procedures like laser eye surgery and breast augmentation be considered moral, immoral, or amoral? Some early Christian thinkers had concerns about altering God's designs: if we try to improve upon the natural world, they reasoned, aren't we implicitly saying that God's creation is *im*perfect?[9]

What do you do when two powerful truth claims come into direct conflict? How do you decide who wins? At a personal level, such conflicts really mess us up because they call into question how we understand the world. This happens whenever we are confronted with the possibility (then reality) that we are wrong about something. But it's bigger. People are not made up of a bunch of trivial facts. Rather, we are characters in a story. Several stories, actually. Part of my story is that I am from the tribe of Lutherans descended from Swiss-German immigrants. I play a role in the ongoing culture and story of America—more specifically, California. My wife and I have a *history*, one that now has four kids. My life has an arc, a purpose, and a vocation. To reduce my life to facts (e.g., I am a professor with a family that lives in Southern California) nowhere near captures the matrix of relationships that make me, me.

Another way to say this is that no matter who you are, you believe certain stories. These narratives first shape, then support, your worldview. By now, you might be saying, "What does any of this have to do with technology and Transhumanism?" This is where the rubber finally hits the road. There is a powerful story that has captivated Christians and non-believers alike. It's a tale that will only grow in magnitude as our world moves deeper into the digital life. Without it, I am not convinced that Transhumanism would have the necessary intellectual fuel to get off the ground. Let me formally introduce you to the Myth of Progress.

9. As you reflect on this set of questions, it may be useful to distinguish the difference between God's creation *pre*-Fall and *post*-Fall, and if that distinction changes how we think about manipulating God's creation to humanity's benefit.

The Myth of Progress

When someone hears the word, "myth," they naturally turn to subjects they learned in high school or college about "mythology." Greek mythology, in particular, comes to mind. It encompasses a large collection of stories about gods and demigods, all of which are essentially colorful tales about great events that demand some sort of explanation beyond the natural world. Of course, part and parcel with these stories is the assumption of the modern reader that these events never actually happened in the way it was described. Sure, the Battle of Troy may have taken place as a historically verifiable event, but certainly not as described in *The Iliad*, where gods and goddesses enter the fray to give warriors supernatural abilities to kill and win glory. The result of studying mythology as a topic is that we essentially throw out the baby with the bath water. Because Greek mythology in particular is a collection of supernatural tales, we tend to think of myths themselves as fundamentally untrue.

I suggest we take care not to go this far. Myths make up our understanding of the world, regardless if they include deities or not. As British moral philosopher, Mary Midgley, writes, "Myths are not lies. Nor are they detached stories. They are imaginative patterns, networks of powerful symbols that suggest particular ways of interpreting the world."[10] Another way of thinking about myths is that they are "life-organizing stories" that describe how people interpret their lives for those things which are of ultimate value.[11] Once we turn our attention to these ultimate values, we can begin to see why a particular life-organizing story can have such profound influence on a society.

Take a popular myth of the mid-twentieth century, affectionately called the American Dream. You perhaps can think of a shorthand version of this myth in just a few seconds. Essentially, the American Dream affords every American citizen the opportunity to have a fulfilling job that pays them enough to afford a comfortable home (always with a two car garage and white picket fence!) and support a family. Work hard, and it's all in your grasp. It may not be true in all cases, sure, but it also isn't categorically false. It is distinctly American because, whether true or not, Americans have been aware that not all national governments encourage the pursuit of private property or wealth accumulation. Generations hold on to this dream

10. Midgley, *Myths We Live By*, 1.

11. Zeigler, *Christian Hope among Rivals*, 34.

as a way to keep hope for themselves and their children. Or, alternatively, consider an equally powerful myth that is told in the single word: karma. Although it is difficult to put forth a reasonable argument that karma (i.e., good things will happen to those who do good, bad things will happen to those who are evil) actually exists, millions of people maintain a certain belief that this is reality *and act in accordance with that belief.*

We're getting closer to a full definition of the Myth of Progress (MoP) by surveying its separate parts. A myth, then, is a life-organizing story. But not only is it the *Myth* of Progress; it's also the Myth of *Progress*. We better take care and think well about "progress" as a concept worthy of its own attention. It might be helpful if we break the term down into three components:

- *Movement*. Progress implies movement. A runner who hears the starting gun go off, yet never leaves her place at the starting line is not necessarily a bad person, but she certainly isn't making any progress.
- *Direction*. Similarly progress can only be had if the runner moves along the prescribed track. By following it, she goes in the right direction. Progress is not haphazard; it is organized toward a final goal.
- *Benefit*. Finally, there must be some end point worth pursuing. This is how we judge the amount of progress that is being accomplished. The runner who is just fifty meters from her goal has made more progress than the one who is one-hundred-fifty meters from the finish line.

All three are vital components.

This view of progress seems rather straightforward, though it's actually quite novel. The ancient Greeks interpreted the past in a different light. They believed that history operated in cycles, with ascendancy followed inevitably by decline. There was no sense that the world would forever advance toward a golden age of enlightenment and happiness. In fact, Greek thought essentially required that any achievement of man would forever be subject to decay; the notion of a culture moving toward the Ideal (whatever that was) in perpetuity was a foreign concept.[12]

The *Myth of Progress* (MoP), by contrast, is the story that emerged out of Modernism that essentially sold the idea that mankind could and would continue to advance as a species—culturally, scientifically, and technologically. Three foundational statements provide us with the MoP's content:

12. Ruse, *Monad to Man*, 21.

1. Progress is inevitable.
2. Progress is natural.
3. Progress is inherently good.

These statements are difficult to challenge because our experience in America today seems to verify these conclusions over and over again. For instance, I would guess that most Americans believe that in twenty years' time, society will have more technological innovations than it does today and people will be able to access more information at faster speeds than ever before. We naturally assume that tomorrow will be filled with scientific advancement and those advancements would, on the whole, better our society. Personally, I'm not exactly thrilled with the idea of going back in time, to outhouses and landlines. I, like many of you, believe that tomorrow's innovations are going to make my life a little easier, a little more fun, a little more productive, and a lot more interesting. The myth is powerful![13]

But our commitment to the Myth of Progress goes deeper. The widespread influence of Darwinism has seeped into humanity's understanding of itself. If evolutionary principles are true, aren't humans always in transition? Aren't they in the process of evolving into something stronger, smarter, and more capable of survival? The Myth of Progress is not only assumed as the unstoppable climb of humanity—it's also nature's way of refining us into something better. If we make the assumption that progress is natural and inevitable—and that the advancements we use have consistently improved our quality of life—then it becomes rather difficult to critique progress as anything *but* good. Notice that we have now imbibed the discussion with moral language. Progress is good, not bad. If it's good, then it is imperative that we do everything we can to facilitate it!

This is the general disposition of technophiles and transhumanists. Society *must* advance because it's our moral obligation to advance it. Under the above rubric (progress is inevitable, natural, and good), the technology industry has outpaced the ethicists who might tap the brakes because the former already claimed the moral high ground. In the 1993 blockbuster film *Jurassic Park*, an interesting conversation takes place between the creator of the dinosaur park, John Hammond, and a mathematician-turned-park-visitor, Ian Malcolm. Hammond attempts to defend the use of genetic technologies to create formerly extinct dinosaurs, and Malcolm

13. For a short but helpful examination of the Myth of Progress, see Burdett, "The Religion of Technology," 131–47.

drives home a simple but lethal ethical point. Referring to the science behind the theme park, Malcolm skewers Hammond, "Your scientists were so preoccupied with whether or not they *could*, they didn't stop to think if they *should*." The debate between these two fictional characters is actually a debate about the third tenet of the Myth of Progress: its inherent goodness. When you talk about something's goodness or badness, you're now in the morality game. It's an ethical problem. How *should* we act? What *should* we do? If the "should" question, the ethical question, does not emerge as a strong partner beside the technologies that are being created, tested, and introduced, the game is up and everyone defaults to the Myth of Progress. We lazily throw up our hands and say, "Well, it's new, and therefore, it *must* be better than the old."

Many transhumanist thinkers, to be fair, have considered a wide range of critiques to their notion of progress, quickly offering philosophical rebuttals that vary in persuasiveness. They have the culture's insatiable thirst for new toys on their side. If a culture accepts the fundamental goodness of progress, which I think it largely has, the battle has been won. Every technological pursuit is now morally justified as a way to bring more good into the world. Sure, something that is blatantly evil—say, weapons of mass destruction—can be rejected by society, but the grayer areas are treated with little resistance. The cultural default position is, "Let's do this." Western society has almost moved to a point where it believes science has "omni-competence."[14] Scientific advancement is available across just about every domain of human life and is revered as the new god; it will create, it will destroy, and nothing can stop it. How is a Christian supposed to react to this deeply inscribed myth without surrendering her faith or resorting to picket lines and protests on the front yard of Google headquarters?

Progress and Human Sin

As Christians, where do we go from here? Talking about foundations can get tedious. We often want the punch line without the set-up. Hopefully, you're tracking at this point, noting where the culture has been exerting pressure on you without you feeling it. It's like the tide that slowly brings in millions of gallons of water, but you just don't notice it's happening. Now that our eyes are open to these forces, the question remains: What do I, as a

14. Omni-competence is a term to describe the general belief that science will eventually solve all of humanity's problems.

Christian, do about it? How can I think about progress and innovation in a way that is godly *and* sensitive to the times we live in?

These questions can and should prompt us to return to the rich resources of Christian theology. It's easy to get intimidated by a term like "theology," so let me set you at ease if you are so inclined. Theology is essentially one thing: the way we talk about God in an organized fashion. The term itself means "God words" or "God talk," and all good theology is utterly grounded in the Bible. Many of us are tempted to think of theology as a solution to problems *back then*; too often, we forget that theology is also for *right now*. The theological tools we can put to use here are substantial. So, let's go out to the woodshed and get the big guns, starting with the Christian understanding of sin.

In short, sin is what separates us from God; it's our rejection of his word. When man and woman rebelled against God's design, it created a great divorce that could only be healed by God himself in Jesus. Romans 5 reminds us that "while we were still sinners, Christ died for us." Sin isn't just an individual thing. It's also communal. *We* sin. We have cultural sin, family sin, personal sin, and the-entire-cosmos-is-sinful sin. Some theologians have been tempted to think of sin as ignorance or a loss of willpower, but this view completely misses the mark. It tempts us to think that if we got a bit smarter, or had a bit more resolve, we would solve all of the brokenness facing our communities. The problem of sin is far more sinister and far more dire. Theologians use the term *original sin* to describe how our innate condition is one of rebellion against God, and this was handed down to us by our parents. It is absolutely unavoidable and leaves us absolutely helpless on our own. Thanks, Mom and Dad.

My point is that a strong view of sin reminds us that we don't have the power to manufacture perfect little lives. There is no chance at a perfect society, perfect family, or perfect church. As I like to repeat over and over again to my students: "Sin gets into *everything*." Even non-physical things are affected: your ability to reason, your creativity, even your intentions. As a Christian, if you hear that this-or-that will usher in a new era of enlightenment or progress without reference to God or sin, a warning siren should go off in your brain because no form of human effort can get rid of the rot. Only Christ can.

Transhumanists are promising a future full of health, intelligence, and bliss. A heaven on earth. But shouldn't all of us—Christians and non-Christians alike—be aware that philosophers and kings have been predicting (and attempting to achieve) some level of utopia since humanity's origins? The Tower of Babel in Genesis 11 was one such instance. The desire of the people who built it was to "make a name for themselves" by employing an innovative technique (brick-making) to manufacture a greatness outside of the blessing of God. Their efforts are thwarted in short order as God came down to confuse the peoples' language, making their construction efforts impossible.

The Enlightenment—which birthed the Myth of Progress—has also shared in this optimistic view of humanity's future prospects.[15] One such prediction came out of the late-nineteenth century, when a school of philosophical thought taught that humanity was about to enter into a golden age of reason, intellect, and advancement. Some of these philosophers had the audacity to predict the place where the utopian society would take flight.[16] Any guesses? Prussia. That's modern-day Germany. It took only about thirty years for the most horrifying war the planet had ever seen to take shape. And just in case World War I didn't knock some sense into these dreams of utopia, just twenty years later we had another World War with even more atrocities. Ground zero? Germany.

Utopias fail because *sin gets into everything*.

I'm not suggesting we give up on our collective efforts to make our countries, communities, and families better and happier places. Of course not. Our call as Christians, however, is to discern what the true source of that happiness is.

If we're supposed to get our happiness from unrestricted consumerism . . .
If we're supposed to find our meaning in wealth, conquest, or greed . . .
If we're supposed to be satisfied with devices, connectivity,
and Amazon Direct . . .

. . . then we're going to fall flat on our faces, wondering why these miracle cures fail to deliver on their promises.

The trajectory of the Christian life is one that follows Christ's obedient walk to the cross—the ultimate expression of love and service. As Jesus bound up the broken, so we too have the opportunity to restore, to heal, to visit, and to forgive. We don't do it to win God's approval. We don't do

15. The Enlightenment was a period of Western intellectual history in the seventeenth and eighteenth centuries marked by advances in philosophy, science, and art. Generally speaking, the society began to reject the authority of government and church institutions and placed an emphasis on reason, knowledge, and philosophical liberalism.

16. The most influential of these philosophers was Georg Hegel, a German thinker whose work continues to be foundational reading in philosophy courses.

it because we think that God will reward us with a better life. We do it because we have been freed from our own sin to enjoy the life God has set out for us. That life—the free life—is one of spontaneous thanksgiving and manifests itself in a genuine love of neighbor. This is what some people call the upside-down values of the cross, where God shows his attention to the meek and the lowly, inviting his disciples to serve rather than be served.

Progress is happening—but God both retains the ability to define what true progress is and how it is to be achieved. Earlier in the chapter, I mentioned that progress can be broken down into three components: movement, direction, and destination. Let's re-orient this discussion through a Christian lens. The movement? God does the work. He often accomplishes his will through his followers, yes, but we should remain crystal clear just who is responsible for any truly desirable end. The direction or path? If God has enabled us to walk beside him as he accomplishes his will, then the path we take is the same path as his son, Jesus. The direction we take is not the quickest route to financial success or personal fame. Rather, our walk is along the *Via Dolorosa*, the Way of the Cross. It is here that Jesus teaches us, instructs us, carries us, and encourages us. Here, we joyfully engage in life oriented toward God's design, the servant life, the life of repentance and grace. Each of these emerge as evidence of the Holy Spirit's presence in our lives and serve as authenticating marks of the Christian journey. The destination? The destination is, quite simply, communion with God. Any secular form of progress is guaranteed to leave us underwhelmed. Riches won't fulfill us. Nor will political power or physical beauty. God has the long game in mind—complete freedom in him. The way of the cross, strangely enough, leads us to God's side in perfect love, freedom, and communion.

The movement of the biblical narrative, after all, is one that points to final and eternal communion with God. The Holy Spirit gives us glimpses of this life while still on earth. We are constantly being molded and shaped into the likeness of Christ through the process of discipleship. Discipleship is just a fancy way of saying that we are learning to become more and more like our teacher every day. We are learning the ways of the Master. It doesn't mean we don't sin anymore, and it certainly doesn't suggest that our days will be happier and our families more functional. But the great gain of the Christian is that our failure is met with God's open embrace. He loves, so we can love. He forgives, so we can forgive. When we screw up either by action or by nature, repentance becomes ground zero for us to learn what grace is, and then, to share it with others. Even more shocking, we receive the benefits of the cross of Christ in spite of the fact that our confession, our cries for repentance, are not without sin either! Jesus Christ makes all things new—he's the one who advocates, who stands beside, who speaks on our behalf, who covers the holes left in our stammering confession before God. Something tells me that a community of repentant believers would be a boatload more joyful then a bunch of people relying on social media and their devices to solve their loneliness problems on the end of another download.

For Christians, hope comes from the outside. It must. And because it does, we can look at technology with clear eyes. We can use our iPads and iPhones knowing that they neither give us all good things nor protect us from evil; these roles are exclusively God's and God's alone.

Discussion Questions

1) Consider how many devices you have in your home right now. Does this provide direct evidence of you or your family's internalization of the Myth of Progress? Or, is there another way to consider the situation?
2) In your social groups, who tends to be more orthodox in their approach to new ideas or innovations? Who tends to challenge the *status quo*?
3) Try to brainstorm instances in Scripture when you think God is unfolding something new, something innovating for his people.
4) What were some of the contributions of the Enlightenment? In what ways has the Enlightenment improved our current societies?

Chapter 4

Freedom and the Body

Scottish Soldier: *Fight against that?! No, we will run, and we will live.*

William Wallace: *Ay, fight and you may die. Run, and you'll live. At least a while. And dying in your beds many years from now, would you be willing to trade all the days from this day to that for one chance, just one chance to come back and tell our enemies that they may take our lives, but they'll never take our freedom!*[1]

MEL GIBSON'S FAMOUS LINES from the 1995 film, *Braveheart*, have made its way into pop culture as a classic battle cry of the oppressed. The scene above is a brief conversation between a Scottish soldier, who surveying the enemy English army surmises that the cause is lost, and William Wallace, a rebel turned hero who rallies the Scottish cause at the Battle of Stirling in 1297. The hinge of the conversation is freedom, perhaps the only thing worth charging a superior army against all odds. Wallace wins the exchange, and the Scots charge.

Nothing quite hits at the heart of the human condition as the thirst for freedom. At their best, the ideals of freedom have liberated billions of people from slavery and extreme poverty. At their worst, they become tired excuses for living a life of utter excess at the expense of someone else. After all, "It's a free country! I can do what I want."

1. *Braveheart* (1995), directed by Mel Gibson.

Of course, as any parent will tell you, freedom without responsibility is more like anarchy than true happiness. St. Paul reminds us that, while "all things are permissible for me, not all things are beneficial" (1 Cor 6:12). The ongoing struggle is to understand the proper relationship between the individual's need to have autonomy and their responsibility to the broader community. Here, as in many other circumstances, definitions mean all the difference in the world. Just what *is* freedom? Jesus himself told his disciples that he came to set people free, reminding them that if "the Son sets you free, you will be free indeed" (John 8:36). Jesus' comments beg the question, "Free to do *what*? Free to *be* what?"

It is precisely at this point that Transhumanism enters the fray. Nothing is more important to the transhumanist movement than the ability to change, alter, manipulate, or augment the human body to an individual's preference.[2] Losing that ability essentially means losing freedom itself. When each person is given the absolute final word on their mind and body augmentations, people have the permission to explore a host of options: third arms, magnetic fingers, solar paneled skin, night vision bionic eyes, and more. A warning: your weird-o-meter is about to tick up a few notches in this chapter!

Any critique of Transhumanism, for better or worse, must seriously evaluate its aims to increase an individual's ability to act on their own behalf, to have genuine autonomy. Otherwise, Christians run the risk of hypocrisy—showering praises upon the principles of freedom but then condemning the people who actually put freedom to use in their own bodies. With these potential consequences in mind, let's start with a simple definition of freedom as a starting point: freedom is the ability to make choices without restraint. This will be subject to change, but at least it will give us a launching point for the rest of our discussion.

Body Issues

I'm going to bring up a topic that no one is ever supposed to bring up in polite company: our bodies. I do it to show how quickly issues about your body descend into issues about your freedom and *vice versa*. I'm also doing this to show how freedom is much more than a concept that exists in our

2. This applies well to the person who is uncritically caught up in a life of digital identity-building. A person can create an online persona from scratch, making themselves to be whatever they find meaningful.

minds; it always has concrete, real-life applications. For transhumanists, these applications almost always impact how we use—and think about—our bodies. So, pick an issue related to your body to test this theory: say, obesity. Do you consider obesity a *social* problem? Let's play this out.

A young man chooses to eat poor foods that, coupled with his sedentary lifestyle, lead to his developing obesity. Regardless of his doctor's concern, he plows ahead with a diet high in sugars and starches—leading to some health complications. The man develops diabetes, among other things, prompting him to begin a regimen of injections and medications to manage his blood sugar, complete with regular doctor check-ups. Meanwhile, a legislator is alarmed at the rise in obesity rates in her state and decides to put forth legislation that that makes it illegal to sell certain sugary foods and drinks in mini-marts, gas stations, and movie theaters. She justifies this legislation by arguing that such eating behaviors cost the taxpayers untold millions of dollars in health and emergency care. How do we think about freedom in an example such as this? Can we control what other people do to their bodies? Should we? Is one person's freedom more important than another's? On one hand, the young man will claim—rightly—that he is "free" to eat what he wants. Nobody wants to be the lunch monitor for all of society, and I suspect there are very few who think it's okay for the government to directly intervene in the affairs of my refrigerator. If someone wants a brownie, they should be able to have a brownie! On the other hand, the consequences of obesity can restrict the freedom of *other* people. Insurance premiums might rise for the healthy because of the actions of the unhealthy, taking more money out of the pocket of people who eat smartly—thus *limiting their options, their freedom,* in other parts of life.

Discussions on freedom and bodily integrity have been, at times, rather tense. Decades-old debates on abortion and euthanasia continue to hit the airwaves and courtrooms, both sides deeply entrenched. News stories about sex slave and human trafficking rings horrify us, and rightfully so. Nothing offends us as deeply as a person who is reduced to a broken, exploited body with no ability to freely choose another way. To become a slave is to lose the capacity to control one's own body! Slaves, both ancient and modern, have been considered property to be bought and sold at the owner's whim. There has always been a deep and profound connection between one's body and their sense of freedom. Slaves were at the disposal of their masters—physically, emotionally, and sexually. They, quite literally, had no control over their own bodies. To be free then was to claim back the ability to think, move, and *be* without absolute submission to a master. You could have your body back if and only if you purchased your freedom. But one thing was certain. If you couldn't pay, your body would never be your own.

Transhumanists seize upon this moral issue as a profoundly simple base from which to argue for unlimited human agency. Transhumanist Anders Sandberg writes, "The right to freedom and life imply a right to one's body. If we have a right to live and be free, but our bodies are not free, then the other rights become irrelevant."[3] Think of bodily rights as a key that opens the door to all other freedoms. Without the key, a house may be pretty to look at, but you're never going to live there in comfort and safety.

Sandberg's got a point. Without possession of your own body, freedom is an almost unthinkable concept. It would be like hearing an employer telling his worker, "You're free to travel to San Francisco, but you can't take your body with you." Yet there is a distinct difference in saying, "My body is my own" and "I can do what I want with my body." The first statement seems to be the centerpiece of universal human rights. It's an absolute must to consider one's most basic possession, one's own body, as sacred. Sure, in religious contexts like marriage, we can offer up our bodies to our betrothed. Or, in the military, warriors can sacrifice their bodies for the safety of the unit. Yet these are willful and thoughtful acts, done without compulsion. In each case, the person never loses possession of his own embodied self in the midst of these sacrificial acts.

However, the second statement above ("I can do what I want with my body") can cause some problems. It seems to disassociate our decisions about bodily life from its effects on the communities we live in. If I am theoretically given x-ray vision through surgery, I most certainly change the community I live in whether I want to or not. Now I have the potential to violate my neighbors' privacy just by looking at them, something that was once completely unattainable and off-limits for me. That potential alone is enough for others to look at me in a different way, to glare in my direction with mistrust, asking, "Is he looking through me right now?" This is important and mind-blowing: your body directly affects your community. Just stew on that for a moment, and we'll return to it soon enough.

3. Sandberg, "Morphological Freedom," 57.

Morphological Freedom

One of the centerpiece doctrines of the transhumanist movement is the defense of *morphological freedom*. This simply means that a person can do whatever they want to their bodies insofar as it does not harm another, nor get in the way of another's bodily freedom. In short, when I say "morphological freedom," think "body transformation." A young woman can decide to put on eyeliner, pierce her nose, get a tattoo on her shoulder, or remove a mole for aesthetic reasons. Or, she could also choose to have a nano-scale (very, very small) shortwave radio placed in her eardrum to help her sleep at night with white noise or to help her work during the day to the music of Beethoven. These changes are borne out of the individual's desire to either look different (aesthetic concerns) or experience the world in a qualitatively different way (functionality).

I imagine most people have no immediate problems with the former "adjustments." I don't freak out when my wife puts on make-up for a Friday night date, by shouting, "What are you doing? You're changing your look!!! Oh, the immorality!" Fashion, exercise, tattoos. These are all minor ways that we change our appearance in order to enhance our physical features, as well as boost our self-image. But are there limits here? Once technology moves interior to our bodies, a small (but not meaningless) boundary has been broken. Such progress has been made before with pacemakers and cochlear implants, but something strikes me as problematic in the second example. Do you see what I see?

Is there any substantive reason for objecting to total morphological freedom (remember, think body transformation) as I've described it so far?

When Five Sense Aren't Enough

Tied closely to the concept of morphological freedom is the increasingly popular movement called *bodyhacking*. Hackers, in the normal sense of the term, are people who overcome software limitations to produce novel or clever ends, sometimes illegally. Bodyhacking, by extension, is the internal use of cutting edge technologies to do much the same: overcome systemic limitations to produce clever, novel, or useful ends directly for the human body.

Imagine a young woman five years into the future. She wants a better, safer, healthier life for herself and so, a few months before, she boldly entered into the bodyhacking movement by having a biometric data chip surgically inserted between two of her ribs. At the most basic level, the chip monitors her vital signs—blood pressure, heart rate, and blood sugar levels. If any of her numbers are off, she receives a text message on her phone, something like:

> *Your blood pressure is a bit high, Katie. Consider going for a walk during your lunch break.*
> *–SmartSystems*

However, her data chip is far more capable than this. It's wirelessly connected to her fully integrated apartment, which is dubbed a "smarthome." If the chip detects a slight increase in blood pressure related to her stressful workplace environment, it sends along information to the house. The home computer system, in turn, immediately prepares the home in response to her attitude and health needs—mood lighting and music to soothe and calm her stress levels, the selection and preparation of a wine bottle or tea kettle, and a slight increase in the home temperature system. The result? She comes home to a holistic environment that understands her without her awareness. It just *knows*. Fifteen minutes after opening the front door, and Katie already feels a bit more like herself, calm and clear-headed.

I don't know about you, but I'd be pretty excited to own a home that automatically adjusts to my moods and rhythms at the relatively low cost of a chip insertion. I would probably have concerns about having my medical information available to a program on the internet, but on the whole, I could be open to this form of passive therapy.

Bodyhacking is essentially a movement that surgically inserts devices into a person's body to perform a wide range of benefits from health to security to aesthetic appeal. These actions are highly unregulated, which makes some people understandably nervous. The bodyhackers themselves, sometimes called "grinders," view themselves as pioneers; the first to embrace the cyborg (mind-machine merge) future that awaits us all.

Before you start imagining dystopian societies, know that there is a wide spectrum of bodyhacking possibilities, some of which you might be relatively comfortable trying. For example, it's not uncommon to meet a person who already unlocks her doors by waving a hand in front of a scanner that can read the rice-sized chip surgically inserted between her thumb and forefinger. I'm not a fan of needles, so this move might be a bit much for me. But for many, this simple procedure has the benefits of added security and ease of use. Traditional keys work just fine for me, thank you, but I'm not sure there's enough here to make a moral objection.

More extreme examples in bodyhacking deal less with practical effects and more so with appearance. Some grinders have chosen to modify their appearance just because they can, using subdermal LED lights on their hands and arms as aesthetic markers. The lights become the tattoos of the new cybernetic age, signaling your ability to enter into specific social groups or perhaps just reflecting your desire to try something novel. If nothing else, these adventurous souls are comfortable with a world that is increasingly turning toward technologies that move *inward*. They are, for all intents and purposes, everyday practitioners of morphological freedom. They are transforming their bodies piece by piece in order to experience new sensations and to gain abilities far exceeding that of an average, unmodified man or woman.

The possibilities are endless. Some grinders are already installing magnets in their fingertips. The inserts give the person the ability to sense magnetic waves. I'm not sure if this will prevent them from using credit cards or not! Some transhumanists have postulated the possibility of more extreme bodyhacking modifications, including the surgical addition of extra eyes, limbs, or fingers—perhaps even the development of wings. Before you laugh, consider how some people have used tattoos as a way to change their whole body's skin color. Piercings, in extreme examples, have done similar appearance overhauls. Would it surprise you at all if an adventurous young person with means says, "Doctor, give me a set of wings!"?

Some of this feels like it's been boosted from the X-Men movies, with every person choosing their preferred "mutation." It's not that far off. There is a deep human desire to stand out as unique and important. I find it helpful to think of bodyhacking as an *empowering* move in both good and bad senses of the word. For those who have been marginalized or pushed to the edges of power, bodyhacking can become a means by which they establish uniqueness and importance. Suddenly, the outcast becomes a trailblazer, a seasoned expert on a frontier where traditional forms of power have yet to make a mark. On the flip side, such newfound power shakes the foundation of traditional forms of community. I imagine that widespread bodyhacking will lead us into deeper tribalism; that is, individuals will augment their bodies in order to gain entry and acceptance into small groups of the similarly transformed, pushing them further away from the community and a common humanity.

There's one more thing I want you to consider. For a transhumanist, morphological freedom is not isolated to making changes to a person's body. Other features can be manipulated as well. "The Transhumanist Declaration," a document laying out some of the values of the movement, makes the following claim:

> We favor morphological freedom—the right to modify and enhance one's body, cognition, and emotions. This freedom includes the right to use or not to use techniques and technologies to extend life, preserve the self through cryonics, uploading, and other means, and to choose further modifications and enhancements.[4]

This is a fascinating statement; it suggests that every feature of our humanity is subject to unlimited tinkering—our body, our thinking, and our emotions. Read between the lines here. At first blush, artificially changing one's emotions or thoughts isn't that big of a deal, since millions of people take Prozac for mood and drink coffee or Red Bull to help them focus. "The Transhumanist Declaration" is saying much more than that. It opens the door for surgeries that fundamentally change how our brains work at the cellular level, *even if we are perfectly healthy to begin with.* If our brain functions were altered to create more bliss-like states, or a chip was inserted into our skulls to direct our neural processes for enhancement purposes (not simply therapeutic), we must consider what this does to our humanity. Remember our foundational question? What makes humans, human? If someone is modifying and enhancing our "body, cognition, and emotions," to what degree does that modification move us out of the realm of *homo sapiens* into something categorically different? At what point in the process must we stop calling it a human and start calling it a cyborg?

The Ethics of Bodyhacking

If these concerns are genuine, then it's time we put some of these proposals to the test. Christians are called to be ethical people, and so it benefits us to use Scripture and sound reasoning to analyze some of the not-so-obvious consequences of morphological freedom, starting with the bodyhacking movement.

How are these modifications different from tattoos, piercings, cosmetic surgeries or even health-oriented operations to install cochlear

4. "Transhumanist Declaration (2012)," 55.

implants or pacemakers? How do ethics work in these cases? Secular ethical frameworks may give us some solid footing. Take the Hippocratic Oath, for example. The Hippocratic Oath, historically used in medical schools as a foundation for proper ethics, is commonly thought to have two basic components: (1) to help, and (2) (at the very least) to do no harm. I would argue that this oath implies that a doctor has the obligation to withhold invasive surgery *on the healthy*. To operate on a normally functioning man or woman would subject the person to unnecessary risk and thus violate the oath.

I am uncomfortable with doctors engaging in transhumanist surgeries that move beyond the therapeutic because such practices are built on a seriously flawed model of understanding human behavior: namely, that a person is (merely) a machine. More than that, performing medical procedures for enhancement purposes infers that the normal functioning person is already *a failed piece* of machinery. Many transhumanists openly acknowledge this belief as a foundational reason for pursuing technological enhancements in the first place. Operating on perfectly healthy bodies as if they are diseased or unwell strikes me as grotesque. This is a common objection that many have for the practice of abortion as birth control. By removing the fetus, the doctor cannot ethically justify his action under the Hippocratic Oath because he would be forced to admit that a pregnancy is a pathological condition to be cured or treated. It's the only way a doctor (who is faithful to the principle to do no harm) can justify the procedure. Such a rationale, in my mind, seems indefensible.

The transhumanist fascination with body enhancement uses the same logic, albeit without the moral baggage of child's life to consider. A doctor, as a starting position, must consider the human body as a machine that needs upgrading or curing before it ever demonstrates a breakdown in its processes. Therefore, the same doctor (in theory) would support giving titanium knee replacements to high school freshman under the justification that: (1) the new knees will be stronger than the biological ones, and (2) you're going to need them replaced eventually anyway, so he'll call it preventative medicine.

The cascade of effects would impact a wide swath of American culture. Collegiate athletics takes a strange turn, where sports competitions are divided into "modified" and "unmodified" distinctions. The job market opens up considerably to those who have more physical and mental abilities than others, and therefore, those who can afford stronger, more flexible bodies and minds would dominate the best paying jobs. Black markets open up overnight to give the daring (with a little extra cash) a surgically provided advantage in their next university application or job interview. These effects by themselves do not necessarily make such body changes wrong. However, communities need to be considering how deep the rabbit hole goes. Once society opens the door for body enhancement, it can never go back.

Okay, let's say the doctors are out—if they take their oath seriously. Does this mean an individual or company can't pursue these ends?

As Christians, how do we react to morphological freedom and bodyhacking? Everything is permitted, not everything is beneficial. A starting point might be the modest reminder that even bodyhacking cannot exempt you from God's love. Adding another limb or inserting a chip does not take you past the gaze of God. Salvation does not rest on whether or not you have a chip in your hand that allows you to pay for your Jamba Juice. We want to be faithful to Scripture and, therefore, we can proclaim with confidence that nothing can separate us from the love of God in Jesus (Rom 8:38–39).

Christians can also say with confidence that God blesses the exploration and discovery of his creation. He neither gave us full maturity nor omniscience at our birth; life is, at the very least, a journey of discovery. Scientific advancements would fall naturally into this conception of God, and such a premise allows for the godly pursuit of truth, goodness, and beauty in science and the liberal arts. Since all truth is ultimately God's truth, we can joyfully affirm the collective efforts of academic disciplines that use a wonderful variety of techniques to ascertain insights into the deepest, most profound human questions.

Everything we do is a reflection of our beliefs—our identity—so while I have the freedom to explore all things, we certainly want to take care that our beliefs align with our actions. Just as many good things when used in excess become dangerous, so we should be able to see how a tool can be turned into a weapon if not handled properly. A hammer drives home nails with little effort; it can also kill a man with one stroke.

Perhaps one way we can think clearly about body transformation and other transhumanist goals is to acknowledge the transactional nature of inventions. For every benefit (real or imagined) that technology offers us, something is being exchanged. A payment is made. In the example of the smarthome above, I didn't really have an issue with a computer chip reading my basic vital signs to create a better aesthetic environment at home. The transaction is small. What am I giving up? In a superficial sense, I suppose I give up my ability to manually choose lighting, music, and temperature—though it would be easy to override the system should I, for some reason, dislike my smarthome's selection. In the broader sense, perhaps I give up a little ground on the notion that only humans are personal agents, since every day computers seem to act more and more like independent thinking beings. Perhaps the transaction is a bit larger upon second glance. After all, I become more childish (i.e., with less responsibility) as my smarthome makes more and more everyday decisions that were once mine. Could I be stunting my own maturation process?

I am not convinced the chip is calling us to question our own biology as human beings, nor does it threaten the specific human dignity I am given by God. If I felt that similar innovations would imply that something is *always* wrong with me and needing adjustment—then red flags should start flying. The default position that the human body is flawed by design runs counter to the biblical narrative where God declares his gift of embodiment as a great *good*. Therefore, it is our responsibility to examine both the tech and the philosophy *behind* the tech to discern its *telos*, its deep purpose. And that's the task. The task of discernment recognizes when our tools force us to make transactions that are too costly. If our technologies undercut our physicality as its *modus operandi*, then we forfeit the dignity of being creatures made by the hand of the Almighty. After all, God offered his very self in Jesus to be born of a woman in the most humble of physical conditions. The Incarnation forces Christians to confront the goodness of creation.

The Age of Excarnation

"Incarnational theology" is used as a catchphrase in many American congregations these days. Many people are asking, what does it mean to be more incarnational? *Incarnation*, at a basic level, means image made flesh. It describes a transition in which an idea or concept becomes a physical reality especially as it relates to the *human* body. The act of incarnation is evident in the Genesis narrative. First, God has a concept or idea in mind, as he states, "Let us make man in our image" (Gen 1:26). Then, through the process of his creative act, he makes manifest that idea into human flesh. Man brought to life, wholly incarnate. Incarnational *theology* is just a way of saying that Christianity is not simply a set of abstract beliefs or principles.

Much more than doctrine, Christianity is built on a God who became flesh and dwelt among us. All of Christianity hinges on the God-man Jesus Christ who died and was resurrected for concrete, real-life communities. Perhaps you have heard people (even Christians) talk about Christianity as if it's *out there*, a set of ethical guidelines and complex theories about God. This misses the mark. Christianity is essentially and irreducibly about Christ, God made flesh. And that, my friends, is the essence of incarnation.

Incarnation is not only the beginning of the story, it is also the climax. The absolute scandal of the Gospels is that an almighty God who identifies as spirit (John 4:24) would lower himself and become man in Jesus. Jesus is brought to us in the flesh as the absolute perfect Incarnation. The earliest Christian heresies, such as Gnosticism, moved away from this central feature of Christian faith, trying to make proper belief about rejecting the body as a corrupt, decaying sack of guts. In the Gnostic model, only by "thinking the right things" or by "knowing the secret information" could one achieve unity with God. The body, at best, was worthless. At worst, it actively fought against a person's efforts to gain salvation.

For every action, there is an equal and opposite reaction; what's true in physics can also be true for theology. If incarnation means a word or idea made flesh, you can probably piece together a working definition for *excarnation*. Excarnation is the movement from flesh to image. For our current discussion, it means to take all that is embodied, fleshy, and tangible and turn it into data. Into quantifiable information. To transform the body into a picture. Pornography is a great example of this transformation. It strips down the complex sexual nature of the individual and converts it into an image of raw desire. The richness of physical embodiment gets reduced to its most barbaric form. This is not an easy concept, so let me offer another way of thinking about excarnation. In the broader sense, excarnation is the reduction of the entire person into a single dimension. Politicians are particularly good at this. They'll twist an opponent's position on a social issue and in an instant reduce them to a farcical caption: "My opponent hates children!" Evidence of excarnation can be more subtle, of course. Google and Amazon have made a fortune by reducing their customers to a series of zeros and ones, turning each client into an algorithm. Why? To better sell blenders and books.

Incarnation, by contrast, makes us three-dimensional. When we are able to consider the complexity and richness of the human person, we open the possibility that a person has internal tensions and contradictions. Each person becomes a unique mix of spirit, emotion, mind, and body, and as such, resists being labeled with too much simplicity. Excarnation is *à la carte*; incarnation is the whole meal.

Excarnation is not just a cultural phenomenon for non-believers. We Christians can lose some of our incarnational character as well, even in the rituals and worship that we hold so dear. Roman Catholic philosopher, Charles Taylor, noted the historical evolution of these practices and described excarnation in this way: Excarnation is "the steady disembodying of spiritual life, so that it is less and less carried in deeply meaningful bodily forms, and lies more and more 'in the head.'"[5] That's deep. He is arguing that our religious practices have slowly turned away from the holistic, face-to-face, tangible Christian way of being in favor of a set of privately held thoughts. Could he be right? Do Christian expressions of faith reside entirely in a person's mind? I find his statement rather convincing, though I would expand on Taylor's quote by adding two extra bits, to read like this:

> Excarnation is the steady disembodying of spiritual life *and life in the community*, so that it is less and less carried in deeply meaningful bodily forms, and lies more and more "in the head" *and in the device* [my emphasis].

The online social network phenomenon, as mentioned earlier in chapter 2, is a major contributing factor to excarnation, as many people now consider participation in community a digital exercise, not a physical one. On top of this, our reliance on our devices is so great that it appears a generation of young Americans feel entirely comfortable spending hour after hour interacting with a four-inch screen to the tremendous expense of losing local familiarity. I may know more about the civil war in Syria thanks to Google and *The Drudge Report*, yes, but I don't have a clue what my neighbor's name is or the fact that he's a cancer survivor and a widower.

Now, let's connect some pieces. Is Transhumanism, with all of its strange and wonderful promises, incarnational or excarnational? In the grand vision of our old friend, Ray Kurzweil, we would be compelled to choose the latter. Remember, Kurzweil wants to upload his consciousness to a machine to live on forever once his organic body has failed. If you think your mind is the only thing that makes you, you—then you are certainly talking about excarnation. You have moved away from the body completely, preferring a singular (and simplistic) way of reducing the human

5. Taylor, *Secular Age*, 771.

person. But others might argue that this is an extreme case because more approachable forms of Transhumanism highlight and improve upon the bodily capacities. They do not seek to leave the body behind, but rather to give it ongoing upgrades to improve its use both today and tomorrow. For these more "moderate" transhumanists, we are still left with a significant problem. It is difficult to articulate any coherent notion of bodily dignity. That's just a painful way of saying, if your body is just a collection of parts to be rearranged and augmented like a set of Legos, then it's nearly impossible to give the body as a whole any real value. There is no framework by which we can say, "This is okay, but that isn't." As a result, many marginalized groups are at risk: those with severe mental disabilities, children, and infants (particularly *in utero*).

I have mentioned Max More before. He is the CEO of *Alcor Life Extension Foundation*, a non-profit organization that freezes dead human bodies (sometimes just the heads) in hopes that future, as-yet-undiscovered technologies might physically resurrect the corpse. A bit macabre, perhaps, but not out of sync with the central philosophies of Transhumanism. I mention this practice because *Alcor* stands at the crossroads of the problem I am trying to address. First, it makes sense that the organization would freeze heads. After all, *Alcor* reasons that the mind cannot exist without the brain, so the brain must be the engine of the mind containing all that makes consciousness possible. Save the brain, and you save the mind. Second, the question of bodily dignity is thrust into the limelight with cryonics and similar questionable practices. Can Max More argue that having containers of frozen heads and bodies (including a two-year-old's) at his Scottsdale facility engenders a *greater* sense of dignity to the recently deceased?

Remember my contention that all use of technology is transactional in nature. There is a trade off between opportunity and cost. For me, the transhumanist obsession with near constant manipulation, augmentation, or transformation leads us down some roads that I find difficult to stomach. The pressing questions (What am I giving up? What does this say about me as a person? How does this sync with Scripture's mandate to love God and neighbor?) often lead to distressing answers. When and where digital technologies lift up the dignity and moral status of the marginalized, *whether the marginalized choose to use the technologies or not*, I can joyfully offer my support as a Christian. Such times are a heartwarming reminder that, yes, the Digital Age has the potential to help me love my neighbor in better ways.

Our bodies affect our communities. Your body can be a visible affirmation of God's creation, or it can undermine the dignity with which you (and your neighbor) were created. What will the outward witness of your body be? How can I affirm the bodily witness of my neighbors? Far more importantly, what was the bodily witness of Jesus Christ in the Gospels?

Conclusion

Considering the topics of this chapter, maybe we can add a little bit to our understanding of human identity. The body is a central feature to the person—not just a flesh casing for the mind or soul. In our bodies, we shape our communities and profess the dignity that comes from being God's creatures. It's not often that you hear of a person's body as theological (i.e., in relation to God) by design, but I think that's the perfect way to think of it. The physical body is a constant reminder of not only God's gracious act of creating us, but also a reminder of our constant spiritual need. More than that, our body bears witness to the Gospel of Christ in profound ways. St. Paul writes, "We always carry around in our body the death of Jesus, so that the life of Jesus may also be revealed in our body. For we who are alive are always being given over to death for Jesus' sake, so that his life may also be revealed in our mortal body" (2 Cor 4:10–11). We're not all-powerful, to be sure, yet Scripture reminds us that our embodiment acts as a reflection of both Good Friday and Easter. Thank goodness we have a body that gives witness to God's work of resurrection!

Discussion Questions

1) Is it reasonable to consider transhumanist body modification as a form of preventative medicine? Would you add restrictions to morphological freedom? What kind?
2) Can you think of other examples where the concepts of freedom and bodily integrity cross paths?
3) Does the online social network movement foster a greater sense of incarnation or excarnation? Could you argue for the other side?
4) In what ways can an OSN foster immaturity? Can it, in other ways, foster *maturity*?

Chapter 5

Digital Sex

The Strange World of Sexuality

THIS IS THE SCARY chapter. Not because we're talking about sex, as if the topic itself is scary. No, it's frightening because here is where you'll notice that Transhumanism is everywhere, particularly in the way society has evolved in its understanding of sex and gender. It's scary because you might have a teenage boy who has a big bullseye on his chest, put there by the titans of Hollywood, the peddlers of pornography, and the cultural movers and shakers who want nothing more than to demolish every form of restraint. It's scary because you might have a teenage girl who is inundated with messages from the beauty industry that seems utterly committed to reminding her of her perceived faults on a daily basis. On top of that, she is told a fiction that sexual relations are the only way to secure the love of a suitor. To resist is to be lonely. To capitulate is to be stamped with a label.

Making things even more complicated, the digital world has spawned new forms of sexual experimentation available to anyone with an internet connection. Virtual reality, cybersex, and gender presenting, oh my! Just about every other day, some life and culture blogsite posts a new article on the latest in artificial intelligence, including ways that AI can be put to use in the sex industry. I told you in the introduction that it's a jungle out there. But do not despair!

This is also a really, really hopeful chapter. Sometimes we're forced to look at a world gone crazy, and it reminds us that the simple pleasure of sex and the beautiful design of our gender is exactly that . . . simple and beautiful. I earnestly believe that Christianity offers us a view of sexuality that is healthy, fun, and full of freedom in the midst of these challenges. At the core of this claim is the fact that our bodies are fundamentally God's. My body is God's. Therefore, my experience of that body is utterly and completely a *gift*. The gift does not lose its preciousness in the presence of other challengers. Like a flawless diamond of great value, the owner does not feel threatened by the existence of the cubic zirconia market. She simply *knows* what she has is precious and worth the price to protect it.

In the last chapter, we talked about morphological freedom, the belief that people have the right to change one's body or emotions to any degree they want in complete freedom. The natural extension of that freedom includes changes to our bodies that directly impact our sexuality, including issues of sexual intercourse, gender, fertility, and birth rates. We are sexual beings. Since sex (at a bare minimum, no pun intended) is an undeniably physical experience, it shouldn't surprise you to find that sexuality, body-modification, and Transhumanism sit at a natural (or, perhaps, unnatural?) intersection. Much of this chapter, then, is going to feel like running full speed with a bowl full of water. We have a lot of ground to cover, and it is likely we are going to make a mess of things. Hopefully you'll be able to glean some useful insights into our hypersexualized culture and find ways to uphold healthy forms of sexual expression.

The State of the Union

Let's start with a general conversation about sex in America today. It's a rip-roaring industry, fueled by two paradoxical impulses. On one hand, Christian parents tell their children that they are at all costs to avoid sex in order to inculcate Judeo-Christian values. Don't touch. Be careful what you wear. God hates it when you have sex before you're married. If you do it with the wrong people or at the wrong times, sex will be a source of shame that will follow you wherever you go! Sure, this is a caricature, but there's a good deal of truth to it. We do such a good job of smashing our children with an anti-sex campaign that they often can't overcome the stigma of sex when they get married. They can't simply flip the switch: It's now okay! God now loves sex! At times, it looks like the only important piece of advice adults have in their arsenal is the desperate plea, "Hey, it's not all about sex!"

On the other hand, billions of dollars are poured into the pharmaceutical industry to amp up our adult sex lives all in an attempt to make men more virile well into their twilight years. We've moved from, "Keep it down, young man" to "Make sure you can keep it up, good sir." Our culture seems to have lost a measured sense of growing older with wisdom, dignity, and a firm grasp of what makes relationships worth holding onto. It's a confusing landscape of messages. In the push for a never-ending sex drive, our youth are also left with the reverse message, "Hey, it's all about sex, after all!"

Once you throw the ingredient of technology into the cultural stew, modern sexuality gets even more perplexing.[1]

Porn, Pills, and Pixels

Pornography continues to be at the forefront of profit-earning industries in the United States, raking in billions of dollars annually. Along with the military industrial complex, science, and gaming, pornography is among most central players in new technological applications. When new innovations (such as the internet) come along, it does not take long for the peddlers of sex to set up shop. As long as humans are interested in sex, there will be an interest in pornography. Though, I warn you, this might not be the case for *trans*-humans. I say that for a couple of reasons. First, I mentioned earlier that some forms of Transhumanism are attempting to modify the pleasure centers of the brain to cut off permanently feelings of anxiety in order to elevate a sense of non-stop bliss. It is unclear whether or not this bliss would make sex altogether uninteresting as a form of pleasure. Perhaps it's even possible to experience sexual pleasure without external aids; in other words, the future might bring with it a form of meditational sex, making the real thing (or, for that matter, pornography itself) an afterthought.

At this point, one might argue that sex is not solely about pleasure, as if that's the only reason one would pursue a sexual partner. It's also done for relational bonding. Sex creates emotional, spiritual, and physical bonds that are substantial goods apart from the pure pleasure of the experience. However, in the transhumanist future that may or may not come to pass, I'm skeptical that people will use sex only for these functional reasons. Transhumanism is deeply concerned about self-fulfillment, not necessarily relational bonding. If relationship bonding is to be had with a partner or

1. For a thoughtful look at modern sexuality and its nuances, I suggest Jeff Mallinson's work on the subject, *Sexy: The Quest for Erotic Virtue in Perplexing Times*.

spouse, will Transhumanism provide better, more fulfilling options than sex?

The ancient Greeks, believe it or not, were in some ways the first transhumanists. They believed that a man was best served to shun the "lower things" that had to do with a man's bodily desires. Food, drink, and sex were appetites that would bring a man down and prevent him from attaining the Ideal. For a Greek, the ideal was a life of practiced reason, restraint, and philosophy. If the choice was between an hour-long contemplation on the meaning of humanity and a bowl of Ben & Jerry's ice cream, the *true man* would choose the former. If only my students were ancient Greeks!

The connection should be clear by now. Transhumanism draws people away from the profound goods of human sexuality toward a view of the person that is data-driven, intellectual, and highly individualistic. The point of Transhumanism is to create a limitless, totally autonomous god in human form, yet sex (in its sexiest and most Christian form) is responsible to the community, fleshy, and an utter demonstration of our need for one another that gets satisfied in a complementary embodied fashion.

No one is quite sure how Transhumanism will affect interests like pornography. For the present, the Digital Age has done nothing to elevate the culture from its base appetites for sexual pleasure. What it has done is create sensory-laden substitutes for embodied sex through the marvels of virtual reality and robotics.

Virtual Sex?

We're just a few short years from photorealism if futurist predictions hold true. *Photorealism* is a term used to describe a man-made image that cannot be distinguished from a real-life environment. Imagine putting on a headset and it's all dark. Then, the cover is removed and you see a bright, crystal clear image of a snowy mountain. If you cannot tell if the scene is digitally produced, or if you are actually looking directly at the mountain, the image is photorealistic. Technology can't produce this type of image yet, but the major players are getting closer every day. In a technological generation or three, we will have more than just photorealism—we'll have videorealistic environments. Videorealism requires an extraordinary amount of processing power to pull off, but once successful, virtual reality will be so immersive that our brains will struggle to determine which reality is actually real! Virtual reality today, while entertaining no doubt, never really causes your brain to think it might be experiencing another reality. The video just isn't fast or clear enough. The viewer retains a distinct break between the real world and the virtual world. However, the next decade will make the break less defined. And potentially more dangerous.

It is a short jump from photo- or video-realistic VR environments to their applications in the sex industry. When programs can be designed for the user to have sexual experiences in any number of scenarios/locations with any number of partners—*and not be able to keep a distinctive break between virtual reality and reality itself*—I'm concerned that the richness of intimate, committed, embodied sex will be lost. The holistic design of God's initial blessing (to be fruitful and multiply) is traded in for a compartmentalized view of the human person. My sexuality, my body, my mind, my spirit, my community—all disconnected, making me a *dis*-integrated person.

A new term has been created to describe sexual preference, one borne completely out of technologies available to those with an internet connection. *Digisexuality* describes the person who restricts all of their sexual activities to digital or virtual environments. Digisexuals do not engage in sexual relationships with any real-life person in any embodied way. This might strike you as ridiculous. I think it makes a great deal of sense myself, though I don't condone it. If we put ourselves in the shoes of a young man or woman who identifies in this way, perhaps it would be easier to conceptualize.

> John is a digital native. He's never known a time when the internet didn't exist. Most of his college classes are entirely online, where his contribution to class comes in the form of discussion board comments, replies, and a few submitted papers. John's an introvert, so he chose a part-time job that could be done from his apartment—a small-time consulting gig that helps computer newbies get in and around the internet. The job pays for the apartment and makes a small, monthly dent in his student loans. He's made a lot of friends, though he only has two or three that he's spent time with in person. He finds rest in a broad assortment of chat rooms, OSNs, and video game platforms.
>
> The idea of dating a real-life woman is terrifying. He's terrible at thinking on the fly; texts and emails at least allow you to edit your comments after you've typed them out. He doesn't understand the social norms of dating at all, considering them a relic of his parent's era. More than that, John worries about the expense of dating a woman. He doesn't make much so the idea of taking

> a young lady out for dinner two nights a week puts stress on his bank account. He's not even sure he wants to open himself up to the emotional vulnerability that comes with a relationship. His parents are divorced and the two "real-life" friends he occasionally sees both had messy endings to their relationships.

The fiction I just created is not exactly rare, and it should stimulate our empathy. More and more young people are experiencing their worlds largely mediated through a screen. Would it be too much of a stretch to say that John (and others like him) would be attracted to the digisexual lifestyle? Virtual reality, in combination with pornography, makes for a compelling no-vulnerability, low-expense way to have a sexual experience. As a culture, we should expect this view to be commonplace in the next fifteen years, and the consequences will be far-reaching and devastating.

Medicated Sex

For those who aren't interested in a future of virtual reality simulations, designer medications seem to be just as attractive. The rise of Viagra and Cialis aren't necessarily a function of the Digital Age, but there is a connection between Western culture's reliance on medication and the transhumanist principles on self-determination. Each seeks to overcome the natural processes of aging and create a person who can fulfill their impulses without limit.

I should be careful here. Western medicine has been an astounding achievement of human ingenuity. Life expectancy today is staggeringly high compared to just a hundred years ago. Infections have been treated, diseases have been cured, and much optimism remains for future research. I do not wish, for a moment, to undercut these gains. Rather, I would like to turn your attention to the nature of the afflictions themselves. It's one thing to prescribe a Z-Pak to treat a bacterial infection; the consequences of not doing so could be dire. It's an entirely different thing to prescribe medicine so that a person can maintain his sexual vigor well into his sixties. Just like the robotic eye experiment in the Introduction, we should be wary of moves that take us past therapeutic uses of treatment to actual human *enhancement*.

Earlier in the book, I argued that treating healthy individuals as defective pieces of machinery was misguided. Here, we have a similar case. The cultural pressure to be a sexual titan as you pass middle age has created this false need to "fix" those who simply don't find sex the end-all, be-all of existence. Transhumanism attempts to create a human being with no limitations. Both the super-longevity strain of H+ and virility drugs make it a goal to eliminate aging and its effects so that every person may forever be young and virile. Lost here is the profound dignity of living a life that has a natural and beautiful trajectory—one that can only be had if it is finite, one that gently travels from the naiveté and discovery of childhood to the experimentation and anxiety of adolescence to the intimate, nuanced relationships and wisdom of adulthood.

We're left with some distressing destinations. The future of sex in America could, in one direction, leave us with a loss of marital fidelity and the long-term blessing of children while chasing unlimited forms of sexual pleasure divorced from procreation and commitment. Every VR system gives people enough to feed their sexual appetites, as real-life embodied relationships feel too boring, too normal, and too uptight. Why should a man choose his wife when he has the opportunity to pursue a video-realistic fantasy scenario with a celebrity or model, sans the emotional commitment? Likewise, can we really blame a woman for seeking virtual companionship when her husband travels on business 150 days a year? The more we disconnect the act of sex from its effect on our communities, the easier time we'll have excusing the above behaviors and the more difficulty we'll have justifying the Christian view where sex plays a crucial role in fostering healthy, committed families.

In the opposite direction, we risk losing the innate beauty of sex in the midst of pursuing other forms of disembodied or manufactured ecstasy. If the promises of great and lasting chemical satisfaction becomes a reality via transhumanist hopes for radical gene therapy or medication (super-well-being), the effect could be a loss of the sexual dimension of humanity altogether. Sex as we know it disappears. It's possible that the human experience devotes itself exclusively to "higher things" like art, music, and philosophy, and that the public view of sex is degraded to be seen merely as an animalistic urge. When this happens, we lose the great gift of sexuality altogether. Digisexualism is just the first step, a dangerous first step that many of our young people are considering at this very moment.

In either case, I think it's reasonable to expect that many people will ultimately opt out of traditional forms of sexual activity made stable by the bond of marriage. Christians will need more than a "Don't have sex!" strategy if they want to be taken seriously in the decades to come.

Gender Issues

Take a personal inventory of the following scenarios. Rank them from 1 (doesn't bother me at all) to 10 (bothers me a lot): How would you feel if . . .

. . . you found out that your spouse played an online fantasy game in which they played as a character of the opposite gender?

. . . your best friend had an online name/handle that suggested the opposite gender (e.g., a man uses the name, *SportyGirl88*)?

. . . you found out your boyfriend/girlfriend was using a misleading name to specifically engage in online conversations with people of the opposite gender in private chat rooms?

The rules are changing for gender, and the digital world isn't making matters any easier.

Earlier in this book, I mentioned the growing number of active gamers in America, somewhere in the vicinity of 180 million active participants. Most games/applications are entirely harmless on the sexual identity front; you can keep playing *Candy Crush* to your hearts content. Other games require the building of an avatar, where the gamer gets to decide who will be hero in the story. This often includes a fully customizable character where you choose: gender, body style, hair, tattoos, and race. No doubt this gives the gamer the ability to customize their experience while also helping them feel like they're immersed in the virtual world they're about to enter.

In such role-playing games (RPGs), it is not uncommon for men gamers to choose female characters as their avatars. Why? For every gamer there might be a unique reason, but here are some possible answers to the question:

- *Functional*: Perhaps the character helps them in some way. For example, a man might choose a rail-thin female avatar because it helps him hide better in-game or makes for a smaller target for enemy arrows or bullets.
- *Experimental*: Some gamers simply want something different, as for many years, the only characters one could play were male. Therefore, a female avatar allows you to play a game from a different, novel angle.
- *Sophomoric*: Many men choose a woman character because he has the ability to create a fantasy girl, usually scantily clad and heavily armed.

But now, researchers are seeing a new demographic emerging. These men and women are using online interactions and gaming platforms for the expressed purpose of gender experimentation. Rather than play the game for the game's sake, some users are playing the game in order to act and be treated as the opposite gender. This happens in OSNs, as well. Sherry Turkle, in her book *Alone Together*, spends a good deal of research on the phenomenon of presenting. *Presenting* is the intentional misrepresentation of a person's gender online as a way to experiment with new forms of relationship-building and gender identity.

Presenting, in some instances, is harmless. In some respects, every real-life social scenario that we engage in can be full of half-truths and white lies. If Jack wants to date Jill, he might, for instance, fudge some facts about the job he has or the countries he's traveled to—all in hopes to impress the lady. I don't find this particularly eerie. Likewise, in the online environment, if a college-aged woman chooses to jump in a game as a male avatar without giving it much thought, no big deal. Often times, male avatars are her only option. She isn't attempting to deceive others in the game.

Complications emerge when deception is the precise reason for the online engagement. Presenting erodes the trustworthiness that is required for healthy communities to function properly. Online predators are obviously a danger to society; we can all agree that everything should be done to punish offenders. But what about the common husband-wife dynamic? If a husband presents as a female to engage in intimate conversations online with others, does this constitute a violation of the marriage vow? It's not just the relationship that is potentially affected. The person doing the presenting, as Sherry Turkle has noted, is more prone to identity confusion. Once the husband logs off, for example, does his real-life gender immediately click back on—or is there some residual after-effects that make this difficult?[2]

Especially vulnerable are pre-teens and teens who have quickly absorbed the fact that online interactions are anonymous—allowing them, in the midst of raging hormones and social anxiety common in high school, to try out new genders *in order to test as real-world options*. While I do not believe that the Digital Age has created transgenderism or the host of new

2. This effect is powerfully presented in the Netflix series, *Black Mirror*. Each episode stands alone as a critique of our digital culture, though many are dark and difficult to watch. In particular, the episode "Striking Vipers" (season 5, episode 1) confronts the complex issues that emerge from gender experimentation in virtual reality or game environments. See Turkle, *Life on the Screen*, 210–32.

gender categories, I think it is entirely plausible that online interactions have made it easier to justify and to legitimize gender ambiguity. It's the perfect environment to practice a new gender, as well as find others who are attempting to do the same.

FutureSex/LoveProblems

Sex is not just something people *do*. Sex is an industry. We are being sold something at every turn, and the sooner we take hold of this fact, the better we can combat the messages that are breaking through the front lines. As I mentioned at the start of this book, an army that expects the ambush is never caught unawares. So look into the near future with me. What does the future of sex look like from a commercial perspective beyond just pornography? First, I think it's safe to say that we can expect a significant rise in the marketing and acceptance of virtual and robotic sexual services. Already in Europe, doll-only brothels have been surfacing.[3] In addition, several companies in North America are developing human-like robots for the sole purpose of sexual pleasure. It won't be long until "doll-only" brothels will be upgraded to "robot-only" establishments that use robotic women complete with a full range of movement and voice capabilities.

Believe it or not, there's a bunch of data out there suggesting that robot-human relationships will become quite common because of the way that *humans* are neurologically hard-wired. For instance:

- From childhood, we grow attached to certain objects that represent comfort, safety, and happiness.
- We regularly ascribe human characteristics to objects. For example, we talk to our cars and argue with our computers as if they were our companions.
- Humans long for affection and will willingly find other sources to find that affection if human options are limited.

For this to work, research suggests, all a robot has to do is convey a sense of empathy and emotion. Of course, a machine cannot have genuine feelings completely of its own making, but that ultimately will not matter. If it looks and acts human, a robot will be able to satisfy the needs of its user and

3. Rodriquez, "Sex-dolls Brothels Open in Spain."

convince the person that it is, in some way, an actual, empathetic, thinking being, a being with which a sexual relationship could be had.[4]

Famous blues guitarist Stevie Ray Vaughan once wrote a song (or so one legend goes) for his guitar, singing, "She's my sweet little thang / She's my pride and joy / She's my sweet little baby / I'm her little loverboy." It doesn't take much to build a relationship with an inanimate object and ascribe real emotional—even romantic—attachment to it. Even though I have described the transhumanist project as one that is inwardly turned toward personal improvement, it appears that human beings are built for relationships to such a degree that devices, computers, and even robots will fittingly serve as companions, friends, and lovers.

While you might recoil at such a thing, spend a moment to consider why a man or woman would choose to own a robot for sexual intimacy or frequent a robot-only brothel. In a sympathetic light, we could imagine a middle-aged man who lost his wife to cancer and has turned to person-robot sex in order to recover some of the feelings of intimacy he lost with his wife's death. Is this enough to qualify the man as a sexual deviant? After all, you could argue that:

- Sex robots/dolls would be safer (no STDs).
- There would be fewer unwanted pregnancies and abortions in society.
- These experiences could be used in therapy to overcome serious intimacy issues.

I'm not trying to win you over to the strange world of robot sex. Not at all. I am, however, trying to evoke a sense of empathy as a starting point for this discussion. If we're more cognizant of the real struggles of real people, we can better lend a Christian witness and critique to the solutions technology is proposing.

There is certainly concern that the rise of robotic or doll sex could lead to serious forms of sexual violence, enacted on the doll precisely because the doll is a doll. To be violent to a robot behind closed doors incurs no apparent threat of consequence, and if human nature holds, when a door is opened to experiment with evil, people will take that opportunity. Yet this shows us an interesting dilemma about *how we think about thinking devices*: Is the robot a being? Or, is the robot a tool? Or, is it both? How we answer

4. Levy, *Love and Sex with Robots*, 105–81.

this question will determine what emerges as normal social behavior and what is rejected as deviant or socially degenerate.

Childbearing and Demographics

A culture is a rich tapestry of practices, artifacts, art, and meaning. When one dynamic changes, the ripple effects can move outward to affect a large swath of other issues. What happens to a community when more and more people are retreating from the world of dating to pursue virtual or robotic partners? You can probably guess. The biggie here is the stunning decline in national birth rates. It's important that we understand that declining birth rates are the result of several forces, certainly not just the increased availability of online sex outlets. One of the contributing factors, however, is the ease with which young men and women can escape from the pressures and potential pain of an embodied relationship. Economist Edward Castronova believes that we are in the midst of a "massive exodus" from the real world to the virtual world.[5] Men especially are moving out of traditional forms of embodied relationships into digisexualism or virtual living. The result? Fewer singles are entering into the dating pool.

In Japan, the situation is dire and well-documented. More young men are living with their parents well into their thirties, and more women are choosing their career over family. The result? A staggeringly low fertility rate that has Japanese industries struggling to cope with the potential loss of fifty-million Japanese by the year 2070. The culture is technologically savvy, so many young people (men, in particular) are turning toward online and virtual options and turning away from the emotional and financial commitment of dating a young woman. Technology is not the single cause of this remarkable cultural shift, but it is certainly helping a generation of Japanese avoid the responsibilities of marriage and child-rearing. In fifty years, Japan will be a remarkably different place, demographically. Only time will tell if the United States will experience a similar fate.[6]

The unintended consequences of the robotic age are starting to percolate in other fields, as well. A short time ago, the robot Sophia was given citizenship status in Saudi Arabia. A publicity stunt, to be sure, but also a

5. Castronova, *Exodus to the Virtual World*, xiv.

6. The early returns are not heartening. The birth rates in the United States, though not nearly as depressing as many countries in Europe, are still at an all-time low. Data can be found at the Centers for Disease Control and Prevention website: www.cdc.gov.

harbinger of things to come. As robots look more like actual human beings, we will treat them more like human beings and ascribe them proportional moral status. In various governments in the world, legislatures are beginning to discuss how to address the issue of robotic legal status, referring to them as "electronic persons."[7] After all, if the intelligence that a machine demonstrates cannot be distinguished between that of humans, they reason, does this mean that actual personhood is achieved in some way? Full circle now. Imagine how this creates further complications between the sex industry and the legal system. Is it possible to sexually assault a robot designed for sexual pleasure?

What about laws concerning marriage? What constitutes cheating in a world where you have a digital liaison that leads to virtual sex without ever leaving your own home and risking a "real" affair? The fact is that legislatures have to deal with complex and evolving issues every day. I'm not suggesting we should reject further advances in medicine or robotics just because it makes life more difficult for a member of Congress. I am, however, suggesting that technology complicates our discussions on the matters of sexuality, and therefore, the church has an opportunity to provide a rich alternative that lifts up sexuality without commoditizing it. Transhumanists will take a more libertarian position. If people want to have sex with robots, it's their prerogative. If people want to use drug therapies for increased virility, so be it. Christians will have the distinct challenge of evaluating the evolving world of sexuality and discerning which pieces are congenial to God's design in Scripture, which pieces are not, and which parts can be explored further in Christian freedom.

Embodied Life in a Digital World

The rich, full life that God offers is a fully embodied one. The creation narrative in Genesis takes pains to describe the earthy, material nature of our bodies. Not only does God use material to create Adam (through the dirt) and Eve (through Adam's rib), he commands man and woman to use their bodies toward the goal of procreation. Based on the Genesis account, it's impossible to arrive at the conclusion that humans are essentially minds and that's it.

The increasingly disembodied forms of sexuality found in Western culture fall squarely into the concept of excarnation I introduced last

7. Bulman, "EU to Vote on Declaring Robots to Be 'Electronic Persons.'"

chapter. When a young woman withdraws from flesh-and-bones interactions with her friends and neighbors for the relative safety of virtual or robotic interactions, she is making an expensive transaction. She has given up a little bit of her embodied character for an image; she has redacted a part of her humanity from the community she lives in. Excarnation tells us a convincing fiction that we are still fully ourselves in the midst of our digital activities because we—when it all comes down to it—are only *minds*. The biblical truth, however, stands in direct contrast to this narrative. We are built to be concrete people in concrete communities, participating in the life of love and limitation in anticipation of the new physical heaven and earth to come.

It's no doubt tempting to think that our communities do not require our bodies. After all, we can do just about anything these days with another person outside of physical presence. I can talk to my wife, laugh with my wife, even "touch" my wife if I have the right hardware even if I am on a trip five states away.[8] This should suggest to us that community no longer requires my physicality. Do you agree?

I would argue that when we engage in face-to-face relationships (sexual or otherwise), our bodies testify a certain honesty about us whether we like it or not. As I've written elsewhere,

> Ultimately, the virtual world forces society to reexamine questions of authenticity and deception, faithfulness and adultery. For more optimistic OSN users, self-representation in any world, digital or real, necessarily requires a certain amount of shape-shifting. After all, is not every single social interaction, digital or otherwise, marked by a certain amount of withholding or experimentation? A young single man, for example, may meet an attractive woman at a bar, and in an attempt to impress her, may stretch the truth about some of his exploits (e.g., where he's traveled, who he knows, how much he earns). Exaggerations, even flat-out mistruths, are not exactly uncommon in embodied attempts at courtship. Yet even in the mistruths of an over-eager suitor, his body cannot be hidden in the embodied exchange; at least one important piece of truth (in this case, what he looks like, his biological sex, and

8. Several technology companies have been developing haptic suits in recent years. More will be said about this innovation in chapter 7. A haptic suit is essentially a skin-tight suit that sends slight electrical charges to the skin to simulate touch in online environments. For example, a gamer might wear a haptic suit to feel warmth or coldness in an in-game experience.

> probably his gender) forces its way into the exchange. No such information necessarily makes itself known in the online world.[9]

I am trying to impress upon you that the body, in many ways, gives us a certain measure of automatic authenticity made available only by face-to-face dealings because holistic human communication includes gestures, spontaneity, and facial expressions. In the future, I suspect that many of our online social networks will be populated with bots. Bots are essentially non-human characters in computer programs. They will have enough artificial intelligence (AI) to act human in its intelligence and speech, yet no actual person will be behind the curtain. You encounter bots, for example, when you call an airline on the phone and you spend the next twenty minutes negotiating with a digital concierge. Many millions of dollars will continue to be poured into bots, all to make the human feel like they are a part of a loving community that has no actual love to offer.

Perhaps a culture oversaturated with online and digital companions will eventually rebel. Perhaps people initially infatuated with novel expressions of sexuality will ultimately shy away from online and robotic forms of romantic and sexual attraction and return to the warmth of human affection. Only time will tell. The call of the Christian is not to complain about "the way things are" or boycott all things digital. The task is to act as agents of restoration and grace to a world that is looking for love in all the wrong places.

Embodiment as Gift

The Christian perspective on sexuality, I believe, has enormous value for a secular worldview that lacks an understanding of grace. The reason why I am hopeful is that we can loudly proclaim that our very embodied existence is, at its essence, a *gift*. Gifts are not earned. Gifts are not paid for by the recipient. They are simply one-sided transactions for the sake of love.

Think about a gift from the perspective of the giver. Say for a moment that you decided to buy your niece a beautiful bracelet, just *because*. You take care to choose the perfect piece of jewelry, complete with all the bells and whistles of good gift packaging and wrapping paper. One evening, when your families are dining together, you hand deliver the gift to your niece, unable to control your beaming smile. She opens it, and to your

9. Oesch, *More than a Pretty Face*, 162–63.

delight (and hers) she immediately puts on the bracelet and shows it to everyone in the house.

Now, imagine that after she leaves the room (in tears of happiness), her father pulls out a checkbook and says, "That was a nice gift. How much do I owe you?" What?! To pay for a gift received offends the nature *of a gift*. You simply want your niece to receive and enjoy with no strings attached. This is a long-winded way of saying that a gift, by its very definition, implies freedom. The bracelet is now the girls and no longer yours—the heart of the giver, then, *desires* that the gift be used and cherished in freedom.

What does this mean for the recipient of the gift? Since the girl knows that such a gift is given in grace and not out of duty or obligation, she instantly knows the gift's worth. The proper response to the gift is both the attitude and desire to use the gift in a way that would honor the giver. Your niece doesn't run to the bathroom and flush it down the toilet, though she certainly has the freedom to do so (if the gift is actually a gift). Rather, she wears it proudly as if to honor both the gift and the person who gave it to her.

Embodiment, especially as it relates to sexuality, can be appreciated in a similar way. God creates us with a body for a bundle of different purposes: procreation, care-taking, gospel proclamation. Yet intrinsically contained in this very gift is freedom, freedom to take part in the rich, full life offered in Jesus (John 10:10). Alternatively, we can also abuse the gift and choose a life of wanton self-destruction. Or, we can insist on paying for the gift and keep a running tally of who owes what to whom. Freedom opens all of these potential paths. While we are given the mandate to love God and neighbor, Christians are not bound to a minute-by-minute script as if there is one and only one godly option for every decision that has to be made. As we will soon see, freedom in the theological sense points us to God and neighbor and, paradoxically, *away from our own desires*.

Well before Transhumanism hit the scene, Pope John Paul II wrote perceptively about the dangers of reducing sexuality to pleasure or efficiency. He wrote:

> Within this same cultural climate [of radical individual freedom], the *body* is so longer perceived as a properly personal reality, a sign and place of relations with others, with God and with the world. It is reduced to pure materiality: it is simply a complex of organs, functions and energies to be used according to the sole criteria of pleasure and efficiency. Consequently, *sexuality* too is depersonalized and exploited: from being the sign, place and language

> of love, that is, of the gift of self and acceptance of another, in all the other's richness as a person, it increasingly becomes the occasion and instrument for self assertion and the selfish satisfaction of personal desires and instincts. Thus the original import of human sexuality is distorted and falsified, and the two meanings, unitive and procreative, inherent in the very nature of the conjugal act, are artificially separated; in this way the marriage union is betrayed and its fruitfulness is subjected to the caprice of the couple. *Procreation* then becomes the "enemy" to be avoided in sexual activity: if it is welcomed, this is only because it expresses a desire, or indeed the intention, to have a child "at all costs," and not because it signifies the complete acceptance of the other and therefore an openness to the richness of life which the child represents.
>
> In the materialistic perspective described so far, *interpersonal relations are seriously impoverished*. The first to be harmed are women, children, the sick or suffering, and the elderly. The criterion of personal dignity—which demands respect, generosity and service—is replaced by the criterion of efficiency, functionality and usefulness: others are considered not for what they "are," but for what they "have, do and produce." This is the supremacy of the strong over the weak.[10]

This rather lengthy excerpt should ring some bells. Pope John Paul II is describes the effects of excarnation! The reduction of God's beautiful, intricate design to a base human impulse attacks the nature of God's gift of sexuality to us and then equally erodes the gift of our bodies that we offer to our spouse. More than that, you can see how the materialistic view of a technology-driven culture can lead us into some dangerous waters where "efficiency, functionality, and usefulness" rule the day at the expense of the weak.

Once we understand our bodies as gift, then, we are invited to ask the question, "How do I make use of this gift?" I receive my body in gratitude, use it to serve and take care of my wife, and take joy in the sexual relationship that such embodiment opens for both of us. Our bodies are given to us in perfect freedom. Likewise, our sexuality is given in freedom. Our response as free men and women in Christ is simple: use the gift accordingly, be thankful, and teach our children to do the same in time.

10. John Paul II, *Gospel of Life*, 42–43.

Discussion Questions

1) In your opinion, what (if anything) is unique about the current cultural obsession with sex?
2) How might a lifestyle of digisexuality be more attractive than traditional embodied relationships? Is it possible to be a digisexual Christian?
3) What are the advantages of using virtual reality platforms in business? Recreation? Education?
4) What other dimension of the human experience would you categorize as gift?

Chapter 6

Civics, Politics, and the Free Person

To this point, I have been offering a few examples of how technologies shape us, from the ways we tinker with our identities in OSNs to the radical changes happening in modern sexuality. I've been concentrating on the internal ways in which technology affects the individual, though not exclusively. Now, I want to turn and address more fully the social implications of digital technology and Transhumanism more directly. It is worthwhile to consider what our communities might look like if our society continues to assume that progress is good and inevitable, just as the Myth of Progress (chapter 3) would have us believe.

With this chapter, I hope to accomplish two tasks. First, I want to offer a few observations into the changing nature of politics and civic life as the Digital Age takes hold. Second, I want to revisit the topic of freedom as it relates to community. I hope that you will find that, contrary to popular opinion, freedom can be found in the great virtues of humility, limitation, and service. We're amping up the theology in this chapter, so if you get lost here or there, stick it out. Read slowly. Think about the points I raise by talking about them with a spouse or friend. In the end, you will have (hopefully) developed a nuanced understanding of the free life we are given in Jesus.

Information Overload

There is one unbreakable rule for sports broadcasters: do not, under any circumstances, talk about politics or religion. For those of you who are sports fans, you can probably can recall a specific instance when this rule was broken and it how it perturbed you. In December 2012, Bob Costas infamously chose to use the halftime of an NFL game to give his personal views on gun control much to the chagrin of the many who tuned in to watch a football game and not a partisan political speech. For some mysterious reason that still eludes the big wigs in Hollywood and at ESPN, people don't enjoy being lectured to by an elite—regardless of which side of the political aisle they fall on (or none at all).

With the advent of Facebook and the New Media, it appears that political discourse in America has forever changed. Now people are subjected to opinions and hot takes from every sphere of human society whether they choose it or not. The noise is simply unavoidable. In years past, global information was relatively difficult to come by and to confirm as accurate. Therefore, entire news teams were dedicated to finding the stories and breaking them to the general public. The major networks chose personalities that looked worthy of the public's trust. They were to be the "the voice of reason." For my parent's generation, the image of Walter Cronkite tearfully relaying the news of President Kennedy's assassination in 1963 remains a vivid sign of the times, as well as an example of associating a news anchor with the story he told. Whether true or not, Cronkite was considered a trusted man in American media, a man who could be relied upon to tell the public what's going on in the world without an obvious political agenda.

The advent and rise of cable news (particularly in the 90s), and then subsequently the addition of the political blogosphere (in the 2000s), manufactured the demand for a twenty-four-hour news cycle. With so many TV hours to fill and so much competition, it shouldn't surprise us that programming soon turned to political discourse fueled by shock stories, scandals, and partisan bickering. Civil discussion simply cannot compete from a ratings standpoint.

Today, we can shoot potshots from the safety of the internet. With a simple Facebook post, we can virtue signal[1] like a professional; letting everyone know our righteous views on gun control, tax legislation, and

1. Virtue signaling is expressing one's opinion on social networks specifically to show others the moral superiority of their own position.

immigration without ever having the discomfort of actually seeing someone react negatively to our pontificating. Even worse, a young blogger can make a name for himself by intentionally writing inflammatory stories with the hopes that the maxim "All publicity is good publicity" holds true. Average Joe Citizen, while no longer at the mercy of the Big Three networks, is now left with a confusing mess of disconnected soundbites, unsure what stories are important and which are just "fake news." The Information Age provides us with news from all corners of the globe, yet it often fails to provide the context in which these stories make sense. It's easy to read a caption; it takes time and dedication to read and absorb an article then make accurate judgments about its content.[2]

Getting the News Out

To put the case as simply as possible, the Digital Age has both positively and negatively affected the way we *do* politics as well as the way we *talk about* politics. In the positive sense, any technologically savvy citizen can generate interest in a social issue that concerns them; they have the means to publicize their work through various OSNs or blog sites. While in past generations citizens were somewhat captive to the elites in regard to what constitutes important news, the average person is no longer just a consumer but can now generate real social change through digital media. In one of many examples, in 2003, South Korea implemented a ban on American-imported beef due to some contaminated samples that were infected with mad cow disease. A few years later, Korea and the United States signed a trade agreement that effectively reopened the Korean market to American beef suppliers. As you might imagine, the Korean public was outraged at the agreement and almost overnight, protests and vigils dominated the political news cycle. Surprisingly, these protests were not the work of organized labor parties or competing interests in the beef market. Rather, teenaged girls organized and mobilized a sizable portion of the dissent using various online social networks (including a boy band fan site) to disseminate information about the various health and political issues related to the agreement. Intense political pressure was brought to bear on the Korean administration, so much so that the formerly popular President Lee Myung-bak nearly lost his power. Korean public policy, in large part, was shaped by a group of concerned, well-connected teenage girls using

2. See Neil Postman's classic, *Amusing Ourselves to Death*.

chat rooms and tenacity as their means to mobilize. This was unthinkable just fifteen years ago.[3]

Another net benefit is the ability to retrieve information. Never before has the accumulated knowledge of the world's cultures been easier to access, at least for those who have an internet connection. Whether one is seeking info about a federal bill regarding wetlands conservation or last week's city council meeting minutes, the access is now there for any and all to see. At least in theory, the average citizen should be, on the whole, more informed about the workings of their local, state, and federal governments. Of course, there's a huge difference between having information readily available and actually *knowing* the information. Take a personal inventory for a moment: Do you consider yourself more informed/engaged in the political world than you were ten years ago? What online tools do you use to access politics and current events?

The overwhelming noise that OSNs generate—even when users attempt to provide helpful information—makes it difficult for genuine listening to take place. Just one look at a random news story's comment section and you will see some of the most vile, hate-filled diatribes one could imagine. When a person is safely cloaked with the anonymity that the internet can provide and buffered from any potential repercussions, civil discourse is often the first casualty. Unfortunately, these bad experiences carry over to the real world. Cyber-bullying can take many different forms. When a person gets steamrolled by self-righteous speeches on Facebook, they often withdraw from the opportunity to talk about social issues in real-life settings unless they are sure that their opinions won't be received (at the very least) negatively. The result is that political perspectives become more private, less subject to another's scrutiny, and therefore, less empathetic to dissenting opinions. Public political discourse begins to resemble trench warfare.

The average person has far and away more access to information than ever before in history. This is a good thing. Unfortunately, it appears that this access has not translated into actual civic smarts. In fact, it may be that the ever-present access to information has made for a dumber, less engaged electorate.[4] Why do I say that? First, the nature of instant information and the widespread availability of smartphones has created a condition in which

3. Shirky, *Cognitive Surplus*, 31–38.

4. For more insight on this phenomenon, see Mark Bauerlein's *The Dumbest Generation*.

no work is required for retrieval of knowledge. In other words, because a young man assumes that if he needs information about anything and his phone will give him that information when he asks for it, he does not *have to learn* anything. It's counterintuitive to think that a device that should make us all smarter is actually keeping us from the pursuit of knowledge.

Second, the success of political ideas depend greatly on the civic conversations that happen *informally*—at watercoolers, pubs, and wine bars. As we have seen in previous chapters, it's much easier to have the discussions on Facebook, where decorum is in short supply and those who see your posts tend to think like you (after all, you control who to accept as your friend). Without civil discourse, which I would argue is best done face to face, we lose the ability to consider well-formed arguments from the other side. Conversation leads to empathy, the process of respectfully encountering and considering an alternate perspective.

OSNs to H+ ASAP

So far in this chapter, we've noted the shifts in the political landscape brought on by the rise of social media and the widespread availability of information. Now, let's turn our attention to Transhumanism and predict what the future might hold for politics and civic engagement. Does the transhumanist future promise a better brand of political discourse? Will it create better citizens, better communities? Before we answer that, let's work through a simple yet revealing thought exercise.

Imagine that a computer software company in the near future, *DC/Tech*, introduces an artificially intelligent personal computer built for the expressed purposes of making information-based decision-making on public policy issues. *DC/Tech* alleges that its program far exceeds the decision-making capacity of a human being because it can process infinitely more data points and reduce hugely complex issues down to effective policy proposals without the potentially "harmful" effects of emotion-driven biases. The programming is designed to maximize the total quality of life for the communities that put it to work. Now, would you be willing to substitute your state's senator for this "thinking" machine on a six-month trial basis? What if the AI turned out to be a huge success?

Far from a flight of fancy, many futurists are predicting the rise and success of robot policy-makers—*precisely because they are not human*. In other words, since robots do not succumb to human biases in judgment, they neither make irrational decisions nor feel the need to descend into mudslinging with other politicians.[5] In theory, if we had an AI governor or senator, certain outcomes would be quite likely:

- No sex or financial scandals. In fact, it's likely such a system could streamline the financial mess of many local government agencies.
- No petty public disagreements that lose sight of important social issues.
- Decisions that are not overrun by emotionalism, but conceived with logical, practical solutions in mind.
- No need for a long campaign season with endless attack ads.

What's the transaction here? What are we saying about robots? About humans? This is all to point out one very simple point: The more we trust computers (and AI) with small tasks, the more we'll allow them to take on larger and larger portions of human life. It will be like having an autopilot button for human decision-making. Let the computer do it; it's more likely to get it right. Using advanced computer programs in this way is also more likely to distance humans from their personal stake and responsibility in social institutions. Wisdom and virtue are fostered when a person is forced to confront a difficult decision that has moral consequences. What happens to human virtue when a decision is shuffled from a moral agent to a computer? Super-citizenry?

Let's briefly recall the three Supers: super-longevity, super-intelligence, and super-well-being. All three of these have implications for the future of our local communities. If superintelligence is ever realized, voting as we know it will be completely overhauled if not eradicated. There are several reasons for this. One, I find it unlikely that a fully augmented person, complete with superintelligence, would want an unenhanced person to vote on any matter of importance. Such a move would be akin to allowing my six-year-old son to take care of my finances, which, while certainly democratic and gracious, would end with his college education money spent on Hot Wheels and Hershey bars. You simply do not allow a person of so "little" intelligence to make decisions that affect all of society. In fact, like the above thought experiment, some people have proposed that human

5. Perhaps you can see a flaw in the argumentation here. Just because a company says it can make an unbiased computer of this sort does not necessarily mean it can remove the bias of the programmers who unwittingly add their influence into the program.

decision-making in government be replaced altogether, arguing that humans are by nature irrational and prone to poor judgment. Their solution? Use robotic artificial intelligence to make large scale, society-shaping decisions at the highest levels of government. RoboPOTUS, anyone?[6]

Super-longevity will create a series of community questions, as well. It is likely that birth rates would plummet in an era of super-longevity. When death and finitude are no longer concerns for the average person, the issues of legacy recede into the background. One no longer needs a child to carry on the family name. In addition, governments will undoubtedly consider laws that reduce the amount of children a couple can choose to have. If no one ever dies, they reason, our planet's resources will soon be under considerable strain from a swelling population and preventative measures will need to be taken.

As for super-well-being, an augmented transhumanist (or post-human, for that matter) has every reason to consider their own happiness as the pinnacle achievement. It is the goal of life, after all. The *telos*. If every impulse of Transhumanism is oriented toward the individual's personal betterment, why would one descend to the concerns of those less fortunate (read: the un-augmented)? History has told us that differences in social class have not led to widespread outbreaks of altruism by those in power. A far more likely scenario is one where the powerful dominates and exploits the poor and weak as a way to keep their own social advantages.

Let me briefly describe another pressure point, one that is brought to bear by the combination of super-longevity and super-well-being. In current societies, there naturally exists a tension between the generation in power (generally older) and the younger, less powerful generation behind it. Tension is released when the old guard peacefully relinquishes their power (e.g., term limits for politicians) or dies out with time. Because the young know their time to take the reins will come, they are less likely to cause too much civil unrest in the present. In a light example, a grunt-level worker in a computer software firm is unlikely to picket the company over the decisions their boss makes, because they know, at some point, their boss will be gone and they can climb the ladder accordingly. This is all to say that the finite-ness of professions *and* people create a necessary pressure release valve that allows individuals to endure difficult times without resorting to aggression. The presence of the Three Supers begs the question, "What if the elites (bosses, senators, CEOs, etc.) never leave?" Super-longevity

6. Linhorst, "Could a Robot Be President?"

necessarily creates deeply entrenched positions of power and influence that resist the natural ebbs and flows of new blood, new ideas, and new people in charge.

Transhumanism will inevitably create a new elite class, as not everyone will jump into the brave new world of body enhancement. It's not difficult to predict the rise of a small, but significant, group of neo-Luddites who retain a view on human nature that rejects direct technological enhancement. They may not be aggressively anti-technology on the whole, but they will certainly object to their fundamental rights being pushed to the side. If this turns out to be the case, and I don't know how it couldn't, the class divisions between those who are technologically modified and those who aren't will be massive. This is one of the central concerns that critics raise about a trans- or post-human future. The intellectual difference between a post-human and a current human will be greater than the difference between a current human and his pet lizard. How much consideration do you give to lizards' rights these days?

In fairness, the situation may not be as fatalistic as I make it out to be. With a little optimism, we could see a different type of society ushered in by H+ and the Three Supers. For instance, the coming of super-intelligence could very well bring about solutions to some of the world's most pressing and impossible problems: poverty, pollution, health, and education. Just as a benevolent aunt might use her wealth to help family members in need, so the transhumanists might also demonstrate limitless altruism to all corners of life and bringing those of all social classes into a better life together. Our cultures and our communities will have to ask if the potential benefits outweigh the concerns.

Though super-longevity in all of its forms might not be embraced by all, certainly some advances should be seen as a collective good without threatening a traditional view of human nature. The potential elimination of cancer, for instance, would be a stunning victory for the human race, guaranteeing a better quality of life for millions. The results of such radical change are difficult to predict. Yet one feature of human behavior is as predictable as death and taxes: add huge amounts of power to human nature—even post-human nature—and bad things are likely to happen.

Perspectives on Power

Power, in its simplest form, is the ability to act. Yet power does not exist in a vacuum, so it is useful to think of it in terms of balance, balance between an individual's agency and the health of the community.

But power is not simply something held by individual people in their quest to get things done. It's also broadly wielded by organizations, corporations, and the ever-growing government bureaucracy. As Christians, we are left to discern our role within the halls of power. Some Christian traditions, such as the Anabaptists, have historically separated themselves from participation in secular forms of government, regarding the potential for spiritual corruption as too great a risk. Other denominations, most notably American Evangelicals, go to great lengths to both influence politicians/elections as well as enter into public political discourse whenever possible. The Moral Majority of the 1980s came about in this way.[7] The tremendously influential theologian, John Calvin (1509–64), in his attempt to create a Christian society in sixteenth century Geneva, provided much of the theological groundwork for such political activism. An obedient servant of God, he reasoned, has the duty to make God's moral law in the Bible the basis for all matters of civic governance.

I'm less interested in the denominational positions for political engagement, though I am sympathetic to many of the reasons for entering (and refraining to enter) the fray. Rather, I'd like for us to consider how power works in relationship to Christian freedom, service, and charity at a base level, then allow this to shape our outward social witness. In other words, let's work from the ground up by starting with individual Christians, then moving toward local communities in which Christians have influence.

Political Freedom and Christian Community

Christians have the obligation to respect and obey their secular authorities (Rom 13). Jesus himself reminds his disciples in Matthew 22 to "render unto Caesar that which is Caesar's." The Christian is not permitted unbounded power, but rather, God allows for secular authorities to maintain

7. The Moral Majority was a political action group founded by Jerry Falwell that promoted socially conservative issues throughout the 1980s. The group was overtly religious and partly responsible for the successful presidential campaigns of Ronald Reagan in 1980 and 1984.

order and restrain evil.[8] So where does this leave us as Christian men and women? First, Christians must recognize that any talk of freedom moving forward must be in concert with talk of the community. There simply is no such thing as individual freedom disconnected from the broader community. Our freedom always bumps against other people in some way, as I mentioned in chapter 4. Theologian Robert Jenson perceptively asks, "What could the content even of saying that I 'freely' chose to arise at seven instead of eight this morning, if I had risen to no one's company and to no one's concern with my action?"[9] Therefore, the best of Christianity is on display when all of our discussions on freedom include the ways in which we experience life *together*.

Second, freedom for the Christian is less about autonomy (I get to do what I want) and more tightly bound to the concepts of liberation and sharing (I have been set free to love my neighbor). Only God exists as a truly free being, utterly independent and relying on nothing. Being bound to Christ's death and resurrection, Christians have been invited to partake in the divine freedom found in the Trinity—to share in the life of God. We have creaturely limitations, sure, but these limitations can be borne in loving communities of faith. Creaturely life is the *pre-condition for sharing to take place*. Seeing others' needs gives us the opportunity to share. Sharing, in turn, is a crucial step into the life of love that the Trinity embodies.

The Freedom of Limitation?

Having unlimited power is not what it's cracked up to be. In fact, I would argue that being a creature—that is, *not* being a god—can be liberating in all sorts of ways. This statement should surprise transhumanists. Limitation—whether physical, mental, or otherwise—is the big, bad wolf for those who would promote human enhancement; it is the proverbial giant to be slayed. The tenets of Transhumanism, and to a lesser degree the philosophical impulses of technology itself, encourage the near unlimited expansion of power and freedom for the individual person. Let's turn our attention to

8. Martin Luther described this feature of God's rule on earth as the left-hand kingdom where law and order are established through social structures like government and family. The right-hand kingdom, by contrast, is where God makes his will known through the gospel and grace.

9. Jenson, *On Thinking the Human*, 42.

these terms to see if the Christian understanding of power and freedom run in sync with the aims of Transhumanism.

We will start with the concept of power. I think the best approach will be to examine a term that often treated as power's direct antonym: submission. At first glance, this should make some sense. Submission might as well be the polar opposite of power. It relinquishes, gives up, surrenders, and otherwise makes one weak. When a martial arts fighter is locked in hand-to-hand combat with a fellow competitor, he does everything he can to avoid a compromising position that would lead to his defeat. If he finds himself unable to escape or if he's at risk of serious injury, he taps his opponent to make clear his submission. The bout is over.

We do not normally think of submission as a direct line to freedom. Usually quite the opposite. To submit is to say, "You're the boss. I'm the servant." A slave submits, not a free man or woman. But submission shouldn't go so far as to imply slavery. We submit to a variety of authorities every single day.

Take a quick inventory of your life and the places where you would say that you have to submit to some authority. In each of these, what does submission actually look like? We unconsciously submit to directives all the time—stopping at a stop light or being screened at an airport, for instance—and so we should be able to recognize that we all fall under authority in a myriad of ways. Is our submission willingly or grudgingly given? Is it to another person or to a group of people?

I would argue that the biblical idea of submission runs counter to the way we most commonly think of the term. To submit is not to lay down like a beaten dog. Rather, proper submission allows a person to stand after relinquishing a burden.

Every morning, six days a week, the President of the United States receives a document outlining the national security concerns of the day, called the President's Daily Brief. It includes reports from the CIA and other intelligence agencies for the purpose of giving POTUS the most up-to-date information they need to make national security decisions. I have no idea what's on that piece of paper, but I'm sure that if I did, I would instantly fall into a panic attack. Threats to peace from all corners of the globe. Resources that have to be fought for. Delicate alliances that require forging. Terrorist organizations that need to be uprooted. At times like these, I thank God that I'm just a professor. People don't live and die because I signed some document; at worst, they just get a failing grade on their final exam.

Our government system—and my role within it as a private citizen—is designed to have representatives make difficult decisions while I carry on with my day. I realize that I am limited and submit to my local government authorities to make those tough calls. As a result of this submission, I don't bear the overwhelming burden that the President does. I sleep in relative peace and quiet, free from the pressure of leading a nation (and the knowledge that half the country at any one time probably thinks I'm doing a horrible job). There can be peace in submission. There can be peace and freedom in *not* knowing the whole picture.

The "submit" verses in Ephesians 5 tend to spark some colorful conversations in the church. Paul implores wives to submit themselves to their husbands (Eph 5:22). For those stuck in the more common understanding of the word, submission becomes the blank check for a husband to demand anything and everything from his wife. Of course, this both ignores the verse immediately prior ("Submit to one another out of reverence for Christ") and loses the essential beauty of Paul's model that comes immediately after the verse in question. Paul is driving home a critical feature of the husband's role—to be utterly sacrificial on account of his wife. The wife can submit to the husband and receive freedom in return, precisely because she can rest in the assurance that his burden-bearing efforts are done as a reflection of Christ's own sacrificial love.

Likewise, Christians need not fear death as an existential end. We submit, and in our submission, we can face death *because Christ died first.* His power was demonstrated in the weakness of the cross becoming utterly submissive to the will of the Father. Knowing that our Father has all things in his care, we (like Jesus) can give up our demand for power and allow God to foster the redemptive life that comes, in part, through submission. The result? You and I live free from the tyranny of sin and death. Free.

Another way to think about freedom through submission and self-giving is illustrated in Matthew 11 when Jesus speaks to his followers in a discourse about yokes and true rest. By taking on the yoke, the mantle, the way of Jesus, who was the perfectly submissive Lamb of God, we paradoxically find that our burden is bearable and rest is there for the taking. I'm not saying that *every* act of submission is one of joy and complete happiness; even Jesus bore his obedience onto the cross. Yet my point remains, the Christian walk is not a walk to gain more power and influence (as the "prosperity gospel" peddlers would try to convince you). Rather, the emptying of our burdens in trust that God Almighty will take them upon himself becomes the starting point of true freedom. We no longer bear the burden of winning our own salvation. We no longer fret about God's disposition to us. We no longer are forced to look to ourselves for all that is good.

Transhumanism does not necessarily require its followers to engorge themselves on power and domination. Perhaps a more conservative form of the movement will encourage self-giving and protection of the weak. In this way, transhumanist principles can be acknowledged as worthwhile insofar as one defers absolute power to God who in turn purposes each person's life as one of submission and service. Only time will tell if this dream becomes reality or remains a pleasant fiction.

Redefining Freedom

In chapter 4, I gave a tentative description of the word freedom. I suggested that its most common meaning was the ability to act without restraint. Now it's time to seriously rework that definition in light of our discoveries above and the testimony of Christian theology. To do this, we'll have to work from the ground up, beginning with what Christians have historically understood as the core reality of all humankind: we are dead to sin. By starting this way, I hope you can see a very simple, straightforward concept. Men and women have no absolute freedom whatsoever, nor can they gain it by their own efforts or inventions. Martin Luther called this the bondage of the will. Only God acts in complete freedom. *Our* freedom is only unrestricted in one sense—we can, at any moment, reject God and his plan for us. Humanity's natural inclination is to assert his own power over and against God. Adam and Eve, if you remember, ate the fruit in part because of the temptation to be like God.

Luther's thought was captured well in a Latin phrase that I encourage you to put to memory. The phrase is *homo incurvatus in se* (pronounced, HO-mo in-curve-AH-toos in say), which essentially means, "man turned inward on himself." Sin does exactly this. It turns a man spiritually inward until he can only see himself. For several years, I had the opportunity to do youth ministry in Oahu, Hawaii. On multiple occasions, I would see older local men walking around town with a severe bend to their backs. In their youth and middle age, these men were pineapple pickers. They labored for years handpicking fruit from the pineapple fields in central Oahu, only to have their backs fuse into massive hunches. A tragic picture, no doubt, but a picture that can be put to good use. Sin has such a crippling effect on our souls. We get so absorbed in our own needs and obsess over our own power that relationships erode around us. How can you see the face of another person when you are too busy navel-gazing?

Dietrich Bonhoeffer, the renowned twentieth-century German theologian, consistently referred to sin as an exercise of self-serving power. The sinful man is one who exploits others, who attempts to be *as God* (Latin: *sicut deus*; pronounced, SEE-koot DAY-oos). The consequence of this sin is the breaking and fragmenting of God's people.[10] The proud ego of the individual cannot stand the idea that he is limited; his ego would see him become his own God at the expense of his neighbor.

Bonhoeffer is on to something here. He's flipping the popular understanding of freedom on its head. To be truly free is not to do whatever you like, as that will inevitably lead to utter loneliness (since you've pushed out God and your neighbor) and despair. Rather, freedom paradoxically comes in submission. God returns to his rightful place *as God* and the individual finally sees that his creaturely life is best borne with others in service and gratitude. It's not freedom-for-oneself; it's freedom-for-the-other. When God says, "it is not good for man to be alone" he is establishing how best to live in the bodies we have, in the communities we have, with the set amount of years we have.

Christians can praise the fact we have bodies that are not unlimited. Being unlimited means that community has no real meaning, since a person seeks all of his fulfillment, his contentedness, even his desires through himself. Physical limitations aren't bad until you buy into the narrative that you are absolutely autonomous. As soon as you do, you essentially take the place of God and must solely rely on your own internal resources for well-being. Biblical freedom does not lead us to the golden prize of autonomy. It leads to a *handing over*. What belongs to God, stays with God. Our limitation rightly recognizes that we do not have to account for solutions to all the problems. God's resources are relied upon, not our own. The by-product of this freeing is contentedness. *Peace.*

The Free, Forgiven Community

Hopefully, by now, you see the character of Christian freedom.

10. Green, *Bonhoeffer*, 48–52.

- Christian freedom is the ability to love your neighbor rather than yourself, made possible only in the love given to us by the cross.
- Christian freedom is the breaking of the chains of one's own ego, where the inward, selfish posture of our soul is now liberated for something greater.
- Christian freedom is the proper orientation to God where *he alone* stands in the center of existence, taking on the burdens of godhood.
- Christian freedom, finally, is found in the Trinitarian God who loves freely in self-giving in order that we, bound to the life of God in Christ, might love our neighbors by sharing in the life that God gives us.

Our bodily limitations serve as boundaries. We can reserve for God that which is God's already. We are no longer attempting to be *sicut deus*, thank God, just dignified creatures in his fold among other creatures. This just might be the best way to start all discussions on politics, community, and life together. If we start with the wrong understanding of freedom, we may just end up with something that looks more like sin and bondage than the actual free living that comes through service, submission, and limitation.

Discussion Questions

1) Does the word "submission" trigger an emotional response for you? Why?
2) In what ways can we use technology to *advance* our understanding of submission and limitation?
3) How can you turn your freedom into a way to serve your neighbor?
4) Identify one or two ways where you are tempted to act as God (*sicut deus*)?

Chapter 7

Work, Play, and Rest

All Work and No Play Makes Jack a Dull . . . Wait, What?

Several weeks ago, my dishwasher went to appliance heaven. I wish that I could tell you exactly what went wrong, but alas, I'm not what you'd call a handyman.[1] All I know is that the electronics stopped working (to choose the wash settings), the racks were falling apart, and the draining system wasn't draining. In past years, I would have called the plumber and that would be that. However, with a bit of encouragement from my wife, I started looking up repair strategies on YouTube. Sure enough, I watched enough do-it-yourself videos that I proceeded to strip down my dishwasher until, after hours of work, I came to a not-so-professional conclusion: we needed a new dishwasher.

After buying the new appliance, I used the same technique (i.e., cheat on YouTube) to install a beautiful, stainless steel marvel of dish-cleaning efficiency. What would've taken a professional plumber less than thirty minutes took me about two hours. I wasn't miffed at my slow pace at all, honestly. In fact, the first thing I said to my wife upon the successful installation was, "Isn't it cool that some dude just decided to show other people how to do this online? Just because?" I had a lot of fun dealing with this

1. To prove this point, when I bought my first house out of college, I called a plumber to fix what I thought was a problem with (you guessed it) my dishwasher. He walked into my kitchen, flipped a prominent switch on my counter and the dishwasher, of course, worked perfectly fine. Palm, meet face.

particular "inconvenience." I'm not sure if I was working or playing or both, to be honest. Everyday living still has its moments of novelty and learning; the simple joys of solving problems provide us with a sense of satisfaction.

In the previous two chapters, we explored the challenges that technology present to our human sexuality and our political communities, respectively. I have stressed that conceiving of these social goods as gifts might be a healthy place to start. To be a sexual creature is to experience the other person's body as a freely given, freely received gift from God. Similarly, communities that begin through procreation then made diverse through association in local cities and towns also act, surprisingly enough, as gifts from God as well. By receiving the Word of God, partaking in the sacraments, and (strangely enough) living a life in submissive freedom, Christians can relinquish the anxiety of having responsibility for things that they cannot control.

Daily life is a gift as well. From the moment we rise in the morning to the moment we collapse at the end of a long day, we are directly experiencing undeserved life and sustenance from our Creator. Often, you might hear the word "mundane" in the negative sense. It's an adjective people ascribe to things that are routine, boring, and lacking in anything resembling joy. Yet the mundane or day-to-day life of the Christian is where God reveals himself to be the God of everything. Which is a more loving expression of parenthood: To give huge hugs after a child performs an act of great achievement (e.g., they receive scholarly honors or sports trophies), or the day-to-day preparation of food and home for a child to mature into adolescence and then adulthood? For every parting of the Red Sea is the hidden daily provision of the quail and manna.

Daily life is a blessing. It is here that the consistent, reliable rhythms of work and play come about. Here, rest saves us between moments of inspiration and offers us a perspective that moves beyond the here-and-now ecstasy of achievements. In this chapter, we'll talk about the influence of technology on how we work, how we play, and how we rest. After all, the vast majority of our lives is caught up in these three pursuits.

The Work Lottery

We've all had jobs that were . . . less than desirable. Hopefully, many of you have also had highly rewarding jobs that provided you with a sense of vocation, where your efforts were realized into a distinctive, positive contribution to society. Yet not everyone gets to play point guard for the Lakers or host their own A&E television show about home remodeling. Is *all* work important? Is *all* work a gift from God?

Imagine you are a single, middle-aged American with an average paying job that provides you with adequate health insurance and a few weeks of vacation every year. One morning, you click on the TV to get your quick cable news fix before work when the following story breaks: In an effort to boost public perception of our nation's government, the President has decided to offer a "work lottery." Twenty-five Americans will be chosen at random to receive an annual salary of $100,000 for the rest of their lives, increasing in time to cover the cost of inflation. The winners do not have to pay taxes, health care is free (and of good quality), and they may move to another part of the country (if it suits them) to make their newfound money go further. There is but one string attached. The winners will not be allowed to enter the workforce for the rest of their lives. Whatever job they currently hold, they must quit immediately. With this withdrawal, the winner is prohibited from contributing anything of substantive value to the broader culture: they cannot donate, educate, or assist in producing artifacts that would be of material or intellectual value for the community. They may *privately* enjoy art, literature, or music—and even create works themselves, but any resulting product can only be done for private satisfaction and must remain entirely closed off from the public. The winners will still be able to enjoy conversations with whomever they choose, insofar as it does not constitute expertise that could lead to a specific benefit for the community.

Now, think about the situation for a few minutes and ask yourself, "Would I want to win the work lottery under these terms?"

No doubt, some of you are saying "Yes!" with absolutely no hesitation. You wonder why *anyone* would tolerate work at all if given the chance to retire at forty. Perhaps your current work environment is terrible and the compensation is not in proportion to the expertise you offer. Or, you may think that you're not producing anything of value in your job right now anyways—so why not get paid for it? Others just want a long, restful vacation. I suppose forty years (or more) of good living would qualify.

Some of you, however, might hesitate. By now, I imagine you know what question I'm going to ask next: What is the transaction? Putting myself into this thought experiment, if I received the work lottery, my first move would be to move out to the great outdoors—Oregon or Wyoming—someplace I could hunt and fish to my heart's content. But if I was honest with myself, I think I'd ultimately lose a sense that I was a contributing member of society . . . and this is no small loss. My life would feel shallow and empty without any higher purpose to put my efforts to good use. I'd hear my father's voice in my brain, encouraging me to give back to society with the gifts I had been given. And he'd be right.

It is not enough to live life for pleasure, at least in the modern sense of the word. Perhaps it kicks against certain American work-hard-so-that-you-can-play-hard mentality. Or, innately, we know that we were put on this earth for much more than enjoying the material pleasures of a cushy life. How we structure the ins and outs of our daily activities *matters*. We want our jobs to matter, we want our relationships to matter, and we want a life that more resembles a deep well than a shallow puddle. When I say "matter," I'm thinking of the Latin word *telos* yet again. It means purpose or direction. If our job doesn't have a sense of purpose, then we question why we put our efforts into it day after day. *Telos* is an essential part of human life. What, or better yet *who*, gives us *telos* is a fundamentally important discovery in the life of the person.

This chapter is all about the *telos* of daily Christian living.

- What is the purpose of our work?
- What is the reason for our play?
- What value is rest?

Hi-ho, Hi-ho, It's Off to Work I Go

It's no secret that the American work force is transforming before our very eyes. Manufacturing jobs that once employed great swaths of Americans in the 50s are now largely automated, driving human labor into the technology and service industries. Many of you remember the days when you had to physically go to a bank and interact with a human teller to deposit your paycheck. Now, smartphone apps take care of that in less time than it takes to read this paragraph. The simple fact is the workforce landscape is a moving target. The skills that were necessary for yesterday's communities often become obsolete as technology inexorably marches forward.

Of course, culture evolves. Just because machines and robots replace human workers in the manufacturing sector doesn't mean that those workers just disappear. Some people are forced to learn new skills. Educational institutions and trade schools have to adapt and use new strategies to gain students; to be relevant, they must look forward and offer their enrollees a useful set of marketable skills. In addition, our changing culture opens new markets and new opportunities for those with the courage to engage them. I say this to prevent us from attacking a straw man. Some of you may want to scream: "Automation is killing our humanity! They're taking our jobs!" While a passionate response without question, I'm not convinced that this is the right way to deal with the issues that innovations create in the workplace.

Transhumanism and the Future of Work

So what happens to the job market ten or twenty years into the future? Let me offer up some potential outcomes that suggest an increasingly technological or transhumanist future.

- *Disembodied or virtual service as commonplace*. It's pretty easy to predict that companies will find new ways to increase profits by lowering costs. In most cases, if a business can implement an online virtual assistant rather than pay for a real person (including wages, health benefits, and vacation time), they're going to do it. Therefore, a steady disembodying of the work force in service jobs is inevitable. Perhaps a better way to put this is to say that human interaction in the public sphere will decrease dramatically. People will always find reasons to gather with friends, but in public settings, where goods and services are exchanged, the most typical interaction will be person-to-computer or person-to-robot. This transformation is already happening and will continue to happen, touching just about every sector of commerce and industry.

  In Japan, sushi restaurants are beginning to automate the entire dining experience. There are no chefs in the kitchen, just a few workers to construct the dishes and keep an eye out for quality control. Nor are there waiters or waitresses as a simple system of conveyor belts run the food past the patron's table. These technologies are even penetrating our own homes. Kitchen robot chefs are already in development and will soon (at an outrageous purchase price) bring a five-star chef right into your own home.

- *Rise of the digital security sector*. Zoltan Istvan, noted Transhumanist and politician, has remarked that in the near future, people will pursue one of two possible job markets: art and cyber-security. He argues that robotic and digital techs will render human labor useless; artificial intelligence supported by robotic hardware will outperform human workers in just about every task. In his mind, this will be a welcome development. No longer tied to physical or intellectual labor, people will have the liberty to use their time advancing the fine arts (painting, drawing, music, and dance) as the human artistic spirit cannot be duplicated by a robot.[2]

  Istvan, however, notes that such a job takeover will require a sizable workforce dedicated to protecting people from external cyber-security threats. As I've mentioned in other chapters, the coming age will require deep thinking across a variety of fields, including ethics and law. A few clever programming bugs after all could grind a flourishing manufacturing plant to a halt in short order. With a society's entire catalog of personal information—from banking to identity-building—in the digital cloud, the need for strict controls and security is a must.

- *Revenge of the analog*. If all of the above outcomes come to fruition, society will develop a small but powerful minority that rejects many, if not most, forms of technology. Some will be motivated by religious reasons, others will call the digital society a fundamentally inhuman enterprise. Regardless, they will gravitate to jobs that require human creativity and craft as a subtle form of cultural protest. Think of handmade wares like guitars or homemade clothing. Some will choose to live "off the grid" in low-density rural areas as a way to retain a humanity that is lost in online communities.

  If human nature indeed requires a sense of embodied, face-to-face experience with others, there is no doubt that the culture will eventually criticize, then act against, the excarnating effect of digital technologies. In 1977–78, the Bee Gees had a nearly unprecedented run of successful songs that swept US and Europe. Three years later, people were burning the albums in public in a massive anti-disco backlash from which the music genre never recovered. Saturate a

2. Istvan made these claims at a lecture he delivered at the Crosswise Institute in 2017 at Concordia University Irvine.

market with one particular product or worldview, and it is just a matter of time before the other shoe drops.

The key here is understanding human nature. Remember that Transhumanism fundamentally rejects any view of human nature that is universal or fixed. If they prove to be right, then society simply evolves like its individuals—without intention or direction. No long-standing resistance to technology will survive. If Transhumanism is wrong and there is in fact an irreducible human nature, then anything that comes as a threat to that nature will eventually stir up resentment and resistance.

Remember, a Christian can look at these developments without resorting to broad-sweeping condemnation. Having a refrigerator that knows your preferences neither isolates you from the love of God nor does it necessarily divorce you from the richness of community. The goal is to see what's coming. The goal is to examine the transaction that's taking place and then live boldly in Christian freedom as agents of restoration in a broken world—no matter where that brokenness is found. The goal is to proclaim a Christian truth about human nature: we are God's children designed to love him, our neighbor, and the broader creation.

The Future of Education

Let's consider a possible future when information downloads become available for every augmented person. If you want to know about the French-Indian War, you simply press a button and your microchipped brain instantly downloads all of the world's available data on the conflict. Such a technology would so profoundly change our society that it is difficult to predict what a culture would actually look like in the future.

We are not quite to the point of direct brain downloads, nor has our society spent enough time discerning if such downloads are a good thing. For the skeptic, this kind of information access might be alarming, but that doesn't necessarily mean technology should be removed from the educational enterprise. In fact, it's important to recognize how digital technologies have improved our educational institutions. More information is available to students *and teachers* than ever before. Best practices can be circulated across hundreds of different platforms, theoretically assisting schools to be better at what they do. In addition, quality instruction and curricula are available to anyone with an internet connection, not just for the rich and privileged. In 2008, Sal Khan began to use short video lessons to help tutor a family member. Now, the Khan Academy is one the world leaders in free online education, teaching everything from higher order mathematics to organic chemistry, all through short YouTube videos. Have you ever learned something from a YouTube or Vimeo video? What is attractive about this style of learning?

Even prestigious traditional universities are opening their resources to the public. Massachusetts Institute of Technology (MIT), in one example, recently opened their entire curriculum (including lecture notes and homework) to the public, completely free of cost. Tuition and fees alone cost the MIT undergraduate about $50,000 a semester, but one can get an MIT education on just about any subject for the cost of a computer and an internet connection.

The present and potential goods here are remarkable. Sure, important arguments need to considered regarding the benefits of an online education *vis-à-vis* a more traditional classroom experience, but in terms of sheer availability to those who want it, free digital education is a home run.

But even a benefit this obvious requires more thought. What if you could take a shortcut around the actual labor of learning? Why spend your precious time fretting over online lessons if a direct download to your brain was possible? Enter Transhumanism. Early in the book, I described Super-Intelligence, a time in which humans have greatly increased their ability to absorb and process information. Perhaps this will come in the form of internal brain microchips that stimulate neural activity or even the availability of direct mind uploads through some yet-to-be-invented technology. Surely this is a great societal good.

Unfortunately, *access* to knowledge and knowledge itself are two different things. Ask any university professor whether or not Wikipedia is a good thing. They're likely to tell you that it is a useful tool in the right hands . . . and they have concerns that students will mistake its presence for actual knowledge. Knowledge alone has never made someone wise. Wisdom happens when knowledge is considered, then usefully applied in a way that promotes the good, the true, and the beautiful. When intelligence is received without the necessary discipline to earn it, a person uses knowledge like a commodity. Proper education in the Digital Age must include not only the information required to make some wise but the actual discipline and discernment to form wisdom in the first place.

Vocation and *Telos*

Essentially, I am arguing that treating work and education as a means to an end is a mistake. It's not that work simply produces income to live or that an education produces raw knowledge. One of the great challenges of our time is to think of our jobs and our education in more holistic fashion. In the Lutheran tradition, the term "vocation" has been given to describe the gift and purpose of work. Whether a person teaches second graders or builds houses for a living, the work they do is a calling, an opportunity to serve God and neighbor through effort. Luther knew the human temptation to look at some professions as better (or more "Christian") than others. In his day, it was largely assumed that God's work was done via the professional ministry class: bishops, priests, monks, etc. A hierarchy of spirituality developed resulted, leaving many common laborers feeling like God treated their lives and their families as an afterthought. The doctrine of vocation reminds us that every job is an opportunity to serve one's neighbor, and therefore, it is intrinsically dignified.

I do not want to give you the impression that if you don't feel directly called by God to do a particular job, that you lack a true vocation. Far from it. Vocation implies a purpose, a purpose for working set forth by God. So, if you have any job that isn't morally repugnant (e.g., being a terrorist), God has invited you into a *telos* for your work. That is, God has given you and your work the purpose to serve your neighbor by the exercising of your unique gifts.

Each one of us is called by God to work. Each one of us has been invited to participate in the bettering of our neighbors by doing that work with diligence and a cheerful spirit. In these efforts, God is given glory and, once again, the gift is received in a manner that is thankful, proper, and good for witnessing the faith.

Avocation and Play

Vocation is a dedicated calling to one's work as service to neighbor. Avocation, by contrast, is one's time *away* from work, spent instead on the pursuits of hobby, craft, or play. For forty hours or more, a person dedicates themself to their job. They must study at a university or trade school to have the necessary skills. They must cultivate relationships in order to be successful at the job interview. Once settled, they work diligently to win a measure of job security, not to mention personal satisfaction. The investment of time is remarkable.

For avocation, the dedication is still present, though for some reason, a person tends to enjoy it more. One of my colleagues, Rod, has the avocation of iron-working. He's a remarkable professor in his everyday vocation, and yet he's convinced that his ironwork on the side contributes something essential to his abilities as a professor and scholar. His ironwork reveals insights to him that can only be experienced through his hands. By sensory touch and hard manual work, the world opens up to him in ways that are inaccessible otherwise. It's a fascinating insight. The body has a certain way to attain knowledge that pure intellectualism knows not. William Gardner speaks to how this might be possible. In his book, *Frames of Mind*, he describes how human cognition is a combination of several different intelligences. One of these specific intelligences is the bodily-kinesthetic intelligence. Essentially, this is a way of knowing that comes from understanding your own body. People with high degrees of this intelligence "have mastery over the motions of their bodies" or are often able "to manipulate objects with finesse."[3] Plenty of activities require this type of learning—woodwork, needlepoint, dance, tennis, etc.

It might be tempting to place vocation and avocation at ends of a spectrum, set in opposition to one another. But, in fact, they are deeply related. Play is actually the purest form of work, and leisure, in its original sense, is the exercising of the mind. We chide our children, "Stop playing around!" Other times, we accuse someone in exasperation, "This is just some big game to you, isn't it?" But playing games isn't a lack of work; it's actually work that is chosen for the simple pleasure of challenge. Any game includes a set of agreed upon obstacles that make achieving the goal difficult. If this wasn't the case, the game wouldn't be fun at all. For tennis to be enjoyable, there must be lines on the court to distinguish "in" balls from "out" balls. A net is erected to make it even more difficult. Once the rules or boundaries are established, then the player works hard to overcome them.

Video games to some extent work the same way. All games require work to solve. But unlike employment, the gamer willingly and joyfully allows for tougher restrictions, harder rules. Only then can the experience of play be a fulfilling one.

Crafts and hobbies create the conditions for achievement through intentional work. There is a deep sense of happiness that comes when a

3. Gardner, *Frames of Mind*, 207.

project is completed. Just ask any craftsman—professional or amateur—who spends hours refining a piece of wood, hammering a piece of metal, laboring over a fine meal, or painting landscapes. The payoff might be the final product, but the joy comes in the process. At some level, people understand the value of working with their hands. Many parents encourage their children to take up physical activities, not just to get them out of the house but to inculcate them with the necessary character that is required to take joy in a time-consuming project.

Some hobbies come through digital means; I do not dispute that fact. Art can come in many forms and graphic design is one of them; many young people are choosing this form of art in either vocational or avocational settings, and for good reason. I am concerned, however, that the rise of digital life (including hobbies) increasingly takes us away from embodied forms of avocation. The community is losing its sense of touch, if you will. Rather than hug, they "like." Rather than dance, they tap their fingers on screens. Rather than experience the bodily-kinesthetic, they lose the physical component of themselves in the device.

Vocation gives a person the ability to contribute to the community. A person can offer a good or service that has societal value. But avocation's value doesn't have to be public at all. Its worth reveals itself in character and virtue. To work at a craft for leisure is to learn about the world by another source: the body.

Avocation is remarkably human. Once again, I want to direct our gaze to the world of Transhumanism. For the transhumanists, the body acts as a flawed piece of hardware. Interesting, but ultimately *in the way*. Their goal is to either: (1) create a super-human body that will overcome the difficulties of our current bodily forms, or (2) transcend the body altogether as a thinking consciousness and nothing more. In either case, the patient virtue of learning about the world through our normal human senses is lost. The character that comes by devoting oneself to craft is cast aside for the instant gratifications offered by digital uploads, body enhancements, and the selfie.

The value of work and the value of play are lost amidst the pursuit of godlike-ness. Yet that's the wrong goal altogether. The ability to do anything and everything without restriction is reserved for God alone. The goal of humanity is better directed toward pursuing what is authentically *human*.

And on the Seventh Day . . .

Rest is foundational to human happiness. As a former youth minister, I can tell you that only two things in this life are truly, truly evil. One is Satan. The other is the junior high lock-in. For those of you who are unfamiliar with a lock-in, the concept is relatively simple: Have parents drop off their son or daughter at the church or youth room around dinner time. Do youth group things (games, movies, bible studies, snacks, more games, etc.). Have the parents pick up their kids at breakfast time. Oh, and no one sleeps.

My first lock-in ended with me trying to sneak in a little catnap at three in the morning, then waking up to find that some boys in the group had a little fun by painting my face with lipstick while I was passed out. Staying up past our limits only leads to bad things. Our patience disappears, our rationality withers, and people get lipstick happy faces written on their cheeks. It's as if our body is shouting, "Don't you realize I can't keep doing this? I need rest NOW!" Without a Sabbath, the vocational life descends into exhausting drudgery.

Does the technological life lead to more or less rest?

The Digital Age has largely succeeded in creating a culture of connection. It's a truly remarkable achievement. Facebook founder, Mark Zuckerberg, on more than one occasion has said that Facebook's primary goal is to connect every person on the planet. With about two billion active monthly users, his vision just may come to pass in its entirety. But maybe, just maybe, being connected to two billion people has created an environment in which a person cannot sit still and be comfortable in their own company.

Connectivity and the Ping

Connectivity can certainly be seen as a good thing. Surely, it is better than total isolation. Still, we would be wise to ask the question, "Is connectivity the same thing as community?" The answer, I believe, is no. At least not any more. Connectivity used to serve as the means to the ends of community. In past years, a woman would pick up a phone to dial a friend (connectivity) in order to set up an opportunity to have lunch (community). Even phone calls that didn't end in embodied contact still provided a measure of relationship-building. Fast forward to today. We are witnessing a generation of young people (and the adults who hypocritically attack them) who use smartphones for the *end* of connectivity. Community is often neither sought nor valued precisely because it's inefficient. It requires human contact, vulnerability, and work.

This brave new world is stimulating, no doubt, as it compels the user to keep an eye on multiple layers of social interaction and sift through countless pieces of fluff. But the situation has moved past a tipping point. The digital generation has moved beyond, "I'm well connected," to, "I'm perpetually *on*." Whether it's the dopamine feedback loops that come from the connected life or it's something more sinister, more than ever, we are losing our ability to separate ourselves from our devices. Our smartphones are always pinging, "Come check this text." Or, "Here's a score update." Or even, "Don't forget that you have a breakfast meeting tomorrow." In many ways, your devices act as a sonar. A submarine sends out discrete pings to determine the relative distances of other objects—the ocean floor, a pod of dolphin, or perhaps enemy submarines. The ping bounces off of those foreign objects and relays this information to the boat. In much the same way, our online social media presence produces a series of pings, all designed to determine our present standing in society. Yet, like the submarine, the individual is isolated from any embodied contact with others. Sure, you notice your friends and acquaintances moving around in the digital world, but to what degree do you *know* them if pictures and posts are all you've got?

Perhaps the question we should be asking is: How can we turn all of this connectivity into forms of meaningful, embodied community?

Human Flourishing

Aristotle thought the best life was a life that pursued happiness. Not happiness as an emotion or attitude, but as an end. To him, *eudaimonia* was human flourishing that manifested itself in an attitude of contemplative reason, virtuous behavior, and worthwhile relationships. And here is where the rubber hits the road. The promised future of Transhumanism most certainly can offer humanity the opportunity for vast contemplation. Many transhumanists predict that our capacity for self-reflection—as an individual, species, even the *cosmos itself*—will grow exponentially with our brain enhancements. The problem is that, unlike Aristotle's vision of *eudaimonia*, the transhumanist vision of human flourishing is disconnected from virtue. You might ask, "Well, can't we find a way to download wisdom and virtue in proportion to our ability to think faster?" Many transhumanists believe this very thing. Yet there are two significant problems with this view. First, a hypothetically "downloadable wisdom" will not be able to stand apart from the transhumanist vision of human flourishing itself. In other words, it would require a transhumanist-friendly attitude to create a downloadable version of wisdom in the first place! The fox is already in the henhouse. For wisdom to flourish, it must be able to evaluate and critique the sources of knowledge it encounters from the *outside*. How can wisdom provide a rational critique against the very philosophy that gave it purchase in the person? It would be as if Steven Spielberg would ask you to critique his movies, saying, "You can say anything you want about the film just as long as you endorse it." Second, wisdom must be practiced. There is no such thing as theoretical wisdom. Wisdom, like theology, is inherently practical. It aligns right thinking with right doing so that personal character might develop through experience.

Let's come full circle. In the transhumanist view of human happiness, virtue is shuffled out the door precisely because it is not earned. When character is downloaded, the recipient might know character at a theoretical level but won't grasp the practical application of it. One could download courage, in theory, but true courage is that which is forged in fire. Only when one practices courageous actions will a man be considered courageous. In this case, there is absolutely no substitute for experience.

Ultimately, enhanced humans may have the opportunity to receive knowledge about a wide, possibly unlimited, array of topics—even topics like virtue and character that lead to human flourishing. However, knowledge is not the same thing as understanding. Understanding is deeper, more profound, and intimately connected to one's personal experience. I know about soldiers. I know what they are, what they do, and how they do it (to a degree). I do not, however, *understand* soldiers. I would have to become one in order to have understanding.

Beginnings and Endings

The beginning must match with the end. From the beginning of this journey, I have said that the only way to have a proper vision of human flourishing is to begin with a proper answer to the following question, "What makes humans, human?" If human beings are essentially machines, which many intelligent people have postulated, human flourishing is a reality where humans run as efficiently as possible. Yet I am promoting a life of vocation, avocation, and Sabbath. These are terms that resist efficiency. These ancient practices are slow; their value in a computerized, data-driven, light-speed world is often obscured from view.

If human beings are essentially animals, then flourishing means to pass on one's genes. But, this, as well, belittles the life of the mind and the contemplation of transcendent ideas like goodness, truth, and beauty.

We have completed our tour through various portions of our shared American culture. I have laid out various issues with regard to our bodies, sexuality, politics, community, work, play, and rest. Now, the time has come to be constructive and lay out a theological way forward (chapter 8) and a structure by which we can understand ourselves in light of our Trinitarian God (chapter 9).

Discussion Questions

1) In the opening thought experiment (The Work Lottery), how did you respond? Is your answer a direct reflection on the level of satisfaction you have at your current job?
2) How does tiredness manifest itself in your personality?
3) What types of activities (other than sleep) give you rest?

Chapter 8

The Resurrection Perspective

I WILL SPEND THE final two chapters articulating a Christian response that is both descriptive (a summation of the problem at hand) and prescriptive (what is the way forward). In this chapter, we will look at the fundamental problem of divine-human separation and see how God's response to our self-imposed distance opens the way for humans to be *human*. This discussion will prep us for the final chapter, where I will put forth a theological attempt to answer the question that has been following us like an obedient dog. That is, "What makes humans, human?" This answer will, I believe, lead us to a position in which we can freely use technologies without becoming slaves to them.

Making the Call

If you have followed a favorite sports team for more than a few years, you have inevitably wanted the head coach or manager fired at one point or another. Perhaps the coach doesn't go for it enough on fourth downs. Maybe he doesn't have his team prepared well for the big games. Or, she's not willing to lay down the law on unruly players. It is as certain as death and taxes. Coaches are always on the hot seat. The win-now mentality of most sports teams is driven by dollars, since better teams attract more fans to attend their games and receive better time slots for prime time games. So much is on the line from a financial point of view that loyalty to any coach, no matter how successful they once were, is now a nostalgic dream of the past.

Still, firing a coach presents several problems. The struggling team will have to adjust to the new system his replacement brings to the table. They will also have to adjust to his management style, since some coaches are known to be player-friendly and others have a more disciplined, tough brand of leadership. A less obvious problem is present as well. What happens if you fire your coach without a viable option for his replacement? This is the great transgression of the over-reacting fan. After a loss, AM radio station phone lines light up like a roman candle with each armchair quarterback shouting, "They need to fire the coach! He's terrible!" Yet, if the radio host is sharp, they might respond with the following question, "Okay, then who would be an available quality replacement?"

It may be low-hanging fruit to criticize one team, one style of play, one philosophy, or one coach, but in order to justify such substantial change, there must be an *alternative* that makes more sense. Pointing your fingers at the problem (diagnosis) is useful, no doubt . . . but it's incomplete. The solution (prescription) to the problem must present itself as well. In an ideal world, the solution addresses the problems well because the new system has a better handle on the reality of the situation than its predecessor did. Back to the coaching metaphor . . . new coaches are successful when they better analyze and respond to the issues of the team than the coach that was just fired.

If a system is failing, a different tack to the situation is a must. It does not do anyone any good to simply stand back, point a finger at the problem, and have a sandwich.

Counteroffers

The same is true for Christian theology when it encounters faulty worldviews. Christians do not advance the gospel by looking at the features of the present culture they do not like and say, "That's a horrible idea!" . . . and end there. Critiques are best when they offer an alternative that better fits reality. For many years, Christians believed that you could prove God's existence by using an argument from absence. They reasoned that since science could not explain all of the natural phenomena in the universe, this lack of knowledge proved that there was a God. In other words, they used God to explain all the negative spaces in our intellectual knowledge. Appropriately so, this apologetic strategy was labeled the "God-of-the-gaps" approach.

Of course, this brand of apologetics was not without its detractors. I'm sure you might be able to identify some of the major problems with this defense of natural theology. One, if the fields of science and knowledge continue to expand (which they almost certainly must), does this mean that God shrinks? Aren't Christians putting themselves on the defensive with each successive scientific discovery? Two, the witness of the New Testament appears to reject the God-of-the-gaps approach, choosing instead to describe the nature of God as revealed in Jesus in real, concrete terms. Jesus himself tells the Pharisees in John 8 that if they knew him, they would know the Father as well. The fullness of God is revealed in and through the tangible Incarnation. Concordantly, St. Paul's ongoing witness in the book of Acts does not defend the Christian God by detailing what he is not—but by what he is! Remember his sermon to the Areopagus in Acts 17? Paul stands before the leaders of Athens and gives this remarkable speech, cutting against Greek polytheism in positive, confident terms:

> Athenians! I see how extremely religious you are in every way. For as I went through the city and looked carefully at the objects of your worship, I found among them an altar with the inscription, "To an unknown god." *What therefore you worship as unknown, this I proclaim to you.* The God who made the world and everything in it, he who is Lord of heaven and earth, does not live in shrines made by human hands, nor is he served by human hands, as though he needed anything, since he himself gives to all mortals life and breath and all things. (Acts 17:22–25)

The twentieth-century theologian Dietrich Bonhoeffer rings many of the same tones as the apostle when he writes,

> How wrong it is to use God as a stop-gap for the incompleteness of our knowledge. If in fact the frontiers of knowledge are being pushed further and further back (and that is bound to be the case), then God is being pushed back with them, and is therefore continually in retreat. We are to find God in what we know, not in what we don't know.[1]

1. Bonhoeffer, *Letters and Papers from Prison*, 311.

The point is essentially the same: it is better to use the *via positiva* and argue for a God that has actual, knowable characteristics rather than to argue for God by what he is not.[2]

Christianity must offer something constructive . . . it can't just chuck rocks from the cheap seats. First, our faith must be a firm proclaimer of who God is as he reveals his character in the Scriptures. The attributes of holiness, eternality, omnipotence, and others permeate the Old and New Testament. We ignore these character traits of God to our own peril. Second, there must be a concurrent commitment to proclaiming the *actions* of God, underscoring how his works have brought about the reconciliation of man through Jesus. These two categories overlap tremendously—to know God's character is to know what he has done and what he continues to do. With this content in hand, Christianity can offer the world an alternative way of living in the Digital Age that better responds to the reality of the human condition with its joys and difficulties.

Design from the Start

My brother Seth is a medical doctor. He has spent years perfecting his craft, a craft dedicated to advancing a set of valuable skills all directed toward one purpose: the health of his patient. If a middle-aged man came into my brother's office and asked, "Am I healthy, doc?" out of the blue, he would not be able to provide a cut-and-dried answer. Seth would no doubt ask him a series of questions about his medical history, his current dietary and exercise habits, and then administer a series of tests. In many ways, his job is to determine a baseline. What is a normal heartrate for this particular man? What is his blood sugar? Without a baseline, Seth would be severely handicapped in his assessments.

Before we get into a discussion on humanity's core sickness, a baseline description would be helpful. Just how far off are we from our designed state? Many writers have mused about life before the Fall, and I have neither the skill nor determination to add anything substantial to that corpus of material. However, I would like to give you a brief look at humanity before the machine went haywire. Scripture gives us a few clues.

2. We have reason to be careful here. Any knowledge that we acquire about God will necessarily be limited and prone to abuse by our sinful condition. Nevertheless, Scripture and the person of Jesus Christ tell us actual, true things about who God is and what he wants for his creation.

- *Created to be together.* After God creates man from the dust of the ground and declares the act very good, he reverses course by stating that man's isolation is "not good" (Gen 2:19). His design is a life that is to be borne *together* with other human beings.
- *Created in freedom to be exercised in freedom.* As God operates in complete and perfect freedom, he freely creates humans out of his love. Part of this creative process includes freedom itself, as man and woman have full agency to operate within the world without compulsion.
- *Created to exercise dominion.* God quickly sets man and woman to a position of leadership over all other creatures. This includes the process of naming, but also includes the responsibility to care for creation and tend to its needs.
- *Created to live without shame or fear.* As any third grader will tell you, Adam and Eve were naked. They knew of no reason to be ashamed as they simply lived in a reality of God's immediate, loving presence.

This is the baseline. God's design for humanity is borne out of love and a desire to share in the life of the Trinity. Rightly ordered relationships are the name of the game: God-to-human, human-to-human, and human-to-nature.

The Problem

If God's will can be discerned through his Word and the prompting of the Holy Spirit, how can a Christian faithfully approach this discussion about technology and Transhumanism? Just about everyone acknowledges that humans are gifted at causing destruction and suffering: They make poor decisions, they regularly hurt other people, and they wreck the environment while they do it. You would have to be a naïve fool to believe that humanity is operating at its highest potential. As the Man in Black says to Princess Buttercup in the timeless classic, *The Princess Bride*: "Life is pain, Highness. Anyone who says differently is selling something." Pain, loss, hardship, suffering. These difficulties are simply part (though not all) of the human condition. Something is keeping humanity from fulfilling its created design. Even worse, this something is more than a condition; it is *actively working against* human efforts to eliminate suffering. Sin, death, and the Devil work in concert to convince humanity that the chains of suffering are unbreakable. Luther knew the power of this unholy trinity, and he knew how quickly they could warp the Christian's sense of their own identity found in Christ.

Common ground exists between Christians and Transhumanists. Both groups loudly profess that the world does not work as it should, though this understanding is packaged in different ways. The Christian sees a broken world because the "should" implies a divine design that humanity has corrupted or simply rejected. The transhumanist sees a world in need of fixing not because of a divine purpose gone wrong but because humanity is shackled to its bodily limitations. I am somewhat heartened by the fact that there is, at the very least, a bit of common ground here. To say that the world requires restoration (as Christians do) is to speak in a similar language with those who would suggest technological solutions to the world's myriad ills. Perhaps Christians can seize the opportunity to build bridges between the competing worldviews in order that, at a bare minimum, we work together to alleviate much of the unnecessary suffering that exists in our communities. Christians have every reason to support research efforts to end scourges like cancer and Alzheimer's Disease; they can work side by side with non-believers (even transhumanists!) to combat the evils of racism and abject poverty. Yet important distinctions remain between these competing worldviews. Distinctions that can make the difference between placing a temporary salve put on society's wounds and enjoying the lasting peace that comes with God's promise of a new heaven and a new earth.

Much of this chapter might already sound familiar to you; I am simply articulating the Christian language of sin and redemption. Hopefully this chapter will aid you as you give further consideration to the state of our world and offer a structure by which you can engage others in constructive conversation. I find it incredibly useful when my Christian brothers and sisters remind me of simple Christian truths. It's like knocking the rust off of some old hinges. When you give the squeaking some attention, your door soon works precisely as it should. Likewise, when a Christian sister draws our attention to the major features of Christian faith and practice, we can more readily diagnose our cultural morass with clear eyes. And here's the kicker: I believe that the power of God's Word can move even the most committed atheist. While words like "sin" and "redemption" might be foreign to our tech-savvy non-believing neighbors, they can still be receptive to the truth of the Scriptural witness through synonyms like "brokenness" and "restoration."

In chapter 3, I introduced the Christian concept of sin. Sin is *the* fundamental problem. You will remember that sin is more than just a collection of bad acts. Sin is systemic. Sin is both individual and collective. Sin is both the committing of wrong acts and the failure to do right. One of my colleagues, a gifted exegete, prefers to translate the Greek word for sin [*hamartia*] as "against God's design." This has been profoundly helpful for me, as it reminds me that God's purposes are always to my benefit. Here I am not so much talking about a minute-to-minute expression of what God wants me to do, as if he has mapped out the only way of holy living amidst the infinite amount of decisions I make with each passing moment. Many Christians get knotted up trying to determine how the minutiae of their lives runs in accordance with a divine gameplan. While I admire this godly commitment, I think the freedom of the Christian gets lost in the weeds with this approach. I am looking at the big stuff here. What does God want for his people? How can I be a conduit to his designs, designs that give us a certain hope for the future? Keeping these overarching themes on our radar reminds us of the blessings he offers to us every day. Conversely, the more we insist on a self-serving definition of "the good" that puts God's ways behind us, the more we find ourselves groping around in the dark.

Sin is what happens outside of God's design. But there's more. Sin not only has a temporal, present effect that erodes individuals and communities. It also brings the final enemy onto the playing field: death.

The Bigger Problem?

In Paul's letter to the Romans, he famously warns that "the wages of sin is death" (Rom 6:23). Or, in my colleague's translation, "the wages of going against God's design is death." Death is the ultimate price for the disobedience in the Garden and the prime object of fear for anything living. Death is the antithesis of God's design precisely because God is the author and sustainer of all life. It is, as N. T. Wright calls it, the "savage break."[3] Funerals are tragic (even Christian funerals) because God's gift of life has been wrenched away, a divorce of the body from the "rich, full life" that God promises (John 10:10). The Deceiver seeks to steal, kill, and destroy. And sometimes, as in the case of bodily death, it appears that Satan has won the game.

3. Wright, *Surprised by Hope*, 14.

What exactly is the individual's most pressing problem? Sin or death? On one hand, you could argue that sin is the core problem. Sin is responsible for all of our relational pain. It is the unholy ground from which springs the worst parts of human nature—lies, greed, lust, envy, and a whole host of other unspeakable deeds. On the other hand, even sinners like us can manage to find happiness in this life. Love is still possible, contentedness still achievable. Death therefore must be the problem. When someone dies, that is the final and complete annihilation of existence. Christians and non-Christians alike cannot tolerate the idea that life will pass from meaning and reality to complete, utter nothingness. According to Transhumanism, the answer would most certainly be death because death is the only term that has non-religious meaning. Earlier in the book, I argue that the super-longevity movement is the lynchpin to all transhumanist thought precisely because the other "powers" that technology afford us are worth nothing when we shuffle off this mortal coil. Death is an enemy to *all* of us, Christian and non-Christian alike.

I argue that both of these terms refer to a more primal issue. The most pressing existential problem is humanity's *separation* from God. Sin is horrible not because I do bad things and that makes my life a little less beautiful. Sin is horrible because it disconnects me from the source of everything good altogether! Death is intolerable not because it is usually preceded by some measure of pain, nor do we avoid death because it stands as the final earthly consequence for Adam's sin. Death is a primal evil because it rends the fabric of life that God has woven for his creatures and threatens to isolate us permanently from his presence.

Solutions

Separation from God—the source of all love, hope, and beauty—is the all-encompassing crisis. The highest good is a life in the presence and communion with the Creator. The beauty of this quest is that it is two-sided. Not only does the Christian desire to be with God in perfect freedom and happiness, God himself directs the entire course of history to bring humanity under his protective fold. Jesus represents the extent to which the Father will seek out and recover his wayward children. Far from a god who looks with disdain or disinterest upon his creation, the Christian God enters into the system that caused the separation and passionately works to repair the damage.

If you asked the average Christian why Jesus had to die, I expect that you would receive a fairly standard answer given with a reasonable amount of confidence. Most, I suspect, would say something like this: "Jesus died in order to save me from my sins" or, "Jesus died as a sacrifice so that my sins would be taken away." I have no issue with the above answers. The confession of the Church from the very beginning has been that Jesus acts as the sacrificial offering for the sins of all people. This is found directly in several portions of Scripture, but perhaps most clearly in 1 John 2 where the author writes: "[Jesus] is the atoning sacrifice for our sins, and not only for ours but also for the sins of the whole world."

If sin and death cause separation between God and man, Jesus (if he actually is the Savior) must confront both. He must defeat sin by being sinless, and he must defeat death by living eternally.

A moment ago, I presented the question, "Why did Jesus have to die?" My hunch was that it was a relatively easy question for Christians to answer. Consider what the answers would be if I followed that question with another: "Why did Jesus rise from the grave?" While I am not certain, I am willing to bet that the range of answers would be considerably greater. I would go so far as to say that many Christians would have a difficult time articulating a biblically-informed response at all. This is a crucial question to which we now turn.

Why the Resurrection Matters

I will not try to offer an apologetic here about the historical veracity of the resurrection. Other authors have put forth excellent texts and cases that are so compelling, that I have no other choice but to joyfully proclaim the resurrection of Jesus as historical fact, worthy of any thinking person's trust. I want to think deeper about the *meaning* behind the resurrection, assuming that the resurrection stands tall to historical scrutiny.

So how, exactly, does the resurrection matter? Doesn't the Bible devote more pages to detailing his life as a rabbi, teacher, and exemplar? I am certainly not belittling the importance of Jesus' earthly life (prior to Easter), but there is a reason that Paul points directly to the resurrection as the cornerstone of the Christian creed. His attention to Easter is so intense that he declares, "if Christ has not been raised, your faith is futile; you are still in your sins" (1 Cor 15:17).

If we can nail down the theological insights that emerge from Jesus' resurrection, we can see clearly how to connect the solution to the actual problem. First and foremost, Christ's resurrection legitimizes his divine claims in the New Testament. This has to be the all-time, single greatest example of "calling your own shot." Jesus predicts his death, then subsequently predicts his resurrection. By doing this, he successfully demonstrates that he and he alone is God. No one apart from a divine force could possibly come back to life after laying dead in a grave over the weekend.

The resurrection of Jesus, therefore, validates his testimony while on earth. All of it. Every person must come to a conclusion about Jesus Christ and his status as God because of the resurrection. While I am not saying that propositions (statements that can be true or false) make up the entirety of the Christian faith, it is true that the entirety of the Christian faith *rests* on the truth or falseness of one proposition: Jesus physically resurrected from the dead. If this proposition is true, then Christianity is true and our entire reality must be oriented to that North Star. If the proposition is false and Christ did not rise from the dead, then Christianity is worse than false. It is not only worth dismissing as untrue; it is worth condemning.

Beyond a vindication of Jesus' divine claims, several other important points must be made. Assuming that the resurrection is factual, we can now direct our gaze to the importance of Easter from a *theological* perspective. Specifically, we need to determine how rising from the dead affects death itself. Let's briefly entertain an alternate history where Jesus dies for the sins of humanity but does not resurrect. We might be tempted to say that his sacrifice for sin is secure and the cost of sin has been paid for with his innocent blood. Even if this were the case, death still stands as the eternal destiny for our physical bodies, as it has not been overcome. But Paul's claim in 1 Cor 15 will not even allow for this form of salvation "lite"—he says that we are "still dead in our sins" if the resurrection is not true. Good Friday can only benefit us if Easter Sunday is connected to it. Scripture speaks in several instances about the nature of death as a rupture between the soul and body. If Jesus does not rise, this break is forever permanent and humanity is never restored to God's design, a design that seamlessly weaves together a perfect union of body and soul to be in communion with himself.

The New Testament claim is that resurrection is the *defeat* of death. N. T. Wright has observed how many Christian leaders have cast aside the true meaning of the resurrection by belittling what death actually entails. They parody the hope of the Christian by retreating to a definition of death that speaks of silence or sleep, trying to mitigate the fear of dying for those who are listening. Wright rightly pounces:

> God's intention is not to let death have its way with us. If the promised final future is simply that immortal souls leave behind their mortal bodies, then death still rules—since that is a description not of the *defeat* of death but simply of death itself, seen from one angle.[4]

As we shall see, the resurrection of Jesus has gutted death of its eternal power. To the Christian, death is still violent and evil, but it is mortally wounded, waiting for its final crushing defeat in God's ultimate Day of Judgment.

Related to death's final defeat, we can confidently proclaim that Jesus' resurrection shows us a picture of the future, a purely good future redeemed by God, not enhanced by some microchip. I enjoy pool time with my children. Not only do I personally enjoy swimming on a hot day, it brings me great joy to see my sons and daughters laugh and splash to their great delight. Perhaps you have seen this common poolside image: a dad stands in the shallow end, attempting to coax his toddler daughter to jump into the water and into his arms. But she's standing on the edge. The edge is certain. Concrete. Knowable. To jump into the water is a leap of faith into the unknown. I have found, however, that if one of her older siblings stands next to her and jumps first, the fear melts away. She sees her own future in her brother's leap and the joy that it brings him. Now the mystery is gone and all that remains is the fun to be had. One, two, three, JUMP!

Jesus has shown us the way through death *and* what lies on the other side. In Paul's letter to the Corinthians, he describes Jesus' resurrection as the template for all those who believe:

> But Christ has indeed been raised from the dead, the firstfruits of those who have fallen asleep. For since death came through a man, the resurrection of the dead comes also through a man. For as in Adam all die, so in Christ all will be made alive. (1 Cor 15:20–22)

A Christian follows Jesus' path in two senses. First, the believer *will actually experience physical resurrection just as Jesus did.* This is no small matter, as resurrection assures each believer that his humanity is complete (more on

4. Wright, *Surprised by Hope*, 15. Author's emphasis.

this in a moment). But, second, Jesus' resurrected body gives us a glimpse into the future where our bodies will be wholly redeemed.

Our resurrected bodies, if they are indeed like Jesus', will be restored into same-but-not-quite-the-same glory. They are the same because they will still be recognizably ours. Jesus' disciples had every reason to be distrustful of their eyes when they encountered the resurrected Lord. After all, who would *expect* to see a friend return to life from the bowels of a sealed tomb? Nevertheless, Jesus showed them the scars that were unique to his body, and those scars serve to reinforce what their eyes were telling them, that this was their Lord! If Jesus can be recognized in his resurrection because of his bodily markers given before his death, so we can expect to have a glorified version of our present form.

Yet our resurrected bodies will be different, as well. Once death is defeated by resurrection, the body cannot die again. We participate in God's new creation without any fear that the divine-human relation will eventually end or die out. Our bodies will discard its perishable qualities for imperishable, and our mortal qualities for immortality (1 Cor 15:53). Jesus' ascension becomes a powerful affirmation of that imperishability in a powerful, albeit short, narrative in the opening chapter of Acts. Here, Luke is accentuating the fact that the risen Jesus did not just live out another fifty years in relative peace to die again. To the contrary, his resurrected body had permanently escaped the clutches of death.

Paul spends a great deal of time linking humanity's fate with the "two Adams." The First Adam sinned, so we all share in the guilt (and consequence) of that sin. But more importantly, the Second Adam (Jesus) overcame that sin as the perfect sacrificial lamb, and because he did, those who believe are bound to Jesus' fate. I can die because Christ died. I will experience the resurrection because Christ resurrected.

The final point I would like to emphasize is that Christ's resurrection opens the way for the restoration of true humanity. What do I mean by this? The Day of the Lord is the culmination of all history. It is the final picture of God, humanity, and nature standing in perfect relation to one another. In some ways you can say that this is when God's design for all of history comes to its full and dramatic conclusion. Every piece is in its precise place—not as a static view of the end, where nothing grows or develops, but rather as a condition of relationship. Humanity is now forever properly oriented toward God as the provider and sustainer, set to worship him in communion with all of creation. Life *after* resurrection *after* death liberates us from bondage, pain, anger, and sorrow. Technology may offer temporary band-aids (manipulating feelings through drug or gene therapies), but it will never come close to resolving the separation that Jesus' resurrection completely cures.

I should be careful here. I am not saying that humanity—even in its current state—lacks the image of God or that it is fundamentally bad. Genesis 9:6 suggests that God's image is still present in humanity even after the carnage of the Fall. It is fair to say, however, that when Adam and Eve rebelled, humanity itself suffered a grave corruption. God's work in history is not to trash humanity and start over from scratch; his work restores us to that perfectly designed condition in the Garden all along. Free, loving, working, playing, resting *humans*.

The hope for the Christian is not a disembodied existence in "heaven" with God. I have been arguing that the creedal hope in "the resurrection of the dead" is central to who we are and what we wait for. True humanity must be embodied. This is what God created on the sixth day. God's design is man and woman most fully alive where body and soul exist together under the lordship of Jesus Christ, the one who made such living possible. Transhumanism would tell you that a human being was destined to become a god, coexisting with other human gods without conflict in perfect harmony. The solution to the suffering of the world by this view is an increase in power made possible by technology so that humans can finally break free from the limitations of their bodies. Yet as we have seen in prior chapters, to be a god is to be completely and utterly alone. More than that, being a transhumanist god not only misses the problem (sin) but it makes the problem infinitely worse (being a god that sins). Using Lord Acton as an inspiration here, "Infinite levels of power corrupt infinitely."[5] The Christian difference is spectacular on this point: the resurrection of Jesus makes humans, human again. Each man and woman is properly oriented toward a God who gives life and exercises authority as Creator while standing side by side with our fellow brothers and sisters to whom we love, care, and bear life in joy together.

That's why the resurrection matters. Jesus' resurrection is the vehicle by which God gives us hearts of flesh and not stone. God recovers and restores humanity's intended design. Women and men can now fulfill their

5. John Dalberg-Acton, a nineteenth-century Member of Parliament, is famously attributed for remarking that "Power tends to corrupt, and absolute power corrupts absolutely."

entire purpose for being, glorifying God as a special part of his physical creation. Our most realized humanity comes not from a clever surgery or a download of information, because neither of these things can fix the heart. Nothing Transhumanism offers can remake a will that is bent by sin and domination. Without Christ, sin will always reign there. Only God transforms. We know he can because he already has. In Jesus, we see our own future. An imperishable body that cannot die. A new heavens and new earth where all of creation proclaims together the wonders of God while performing their highest *telos* as creatures. A new heart and will that puts aside the superficial desires of power, beauty, and intelligence and puts on the infinitely more valuable fruits of the Spirit.

Discussion Questions

1) Could transhumanists make a reasonable argument that super-longevity makes the Christian notion of the resurrection obsolete?
2) How do you envision the coming resurrection of the dead? What sights, sounds, or tactile experiences do you imagine?
3) Do you find it easier to describe God by what he is, or by what he is not?
4) What individual liberties would you being willing to give up if relinquishing them meant bringing super-longevity, super-intelligence, and super-well-being into reality?

Chapter 9

The Christian Person

Past, Future, Present?

The resurrection of Jesus is the single most important event in the history of the created universe. It stands as God's great affirmation of the created world and successfully demonstrates his commitment to bring all of life back into a joyful, healing relationship. I treat the empty tomb as a type of skeleton key, a tool that opens all other doors. Because Jesus was resurrected, the Christian sees the present world not as a doomed planet of domination and environmental destruction, but rather as the place where God will make all things new. Christians are given the task to work as agents of restoration in anticipation of God's final redemptive work. Because Jesus was resurrected, the Christian sees their own life not as an opportunity to get as much as they can before they die, but rather as an instrument of God's design where they joyfully fulfill their vocations on both sides of death. Resurrection turns death into life, decay into opportunity, and fatalism into hope.

At the end of the last chapter, I argued that our hope is bound to the events of Easter morning two thousand years ago. One way you could say this is that a past event has given us a glimpse—perhaps even a foretaste—of the future. Past, future. That leaves one tense left.

If resurrection is the hope to which the Christian clings, what happens in the *now*? Perhaps you've had moments where you lose yourself in a daydream. Your vacation week is only four days away. You may be at work physically, but in your mind, you are hiking and fishing—or reading a stimulating book on theology and technology. Living too much in the future puts your present at considerable risk, just as daydreaming while driving can quickly lead to a fender bender. It is best to show a little restraint. To properly think of hope, then, is to think of it as an overarching condition that penetrates everything. It is much more than a wished-for future compartmentalized as a something to be experienced down the line at some unforeseen time. Hope helps set the trajectory of one's *present* life. If I place my hope in a future stock market rebound, I am not just wishing it to come about out of nowhere. Rather, I make plans in the present to be fully prepared so that I might take advantage of hope becoming reality. Yet even this example is lacking, as it implies that hope is restricted to a "becoming" thing. I suggest that hope *makes* reality for the Christian. It is the certainty of things to come so that the Christian can see God more clearly in all of their affairs—their daily joy, their suffering, and their purpose. Yes, we wait for the full consummation of God's plan. This is a part of hope. But let's not lose sight of the fact that those who have Christian hope put their efforts to aligning their present with the future that is promised in Jesus.

We have covered quite a bit of ground. We are now at a place where we can consider a theologically grounded understanding of our unique human-ness. Our technological society has offered a materialist view of the person and its hope that society will find ever more creative ways to shield its citizens' eyes from the reality of suffering and death. For the transhumanist, the uniqueness of humanity lies not in its essential nature or created dignity but in its ability to defy its own physical limitations through applied reason. Humanity can and should evolve itself. Hope can be found if and when people are able to wield technology against death, giving them the unlimited freedom to pursue their ends with little or no restriction. If you pay attention closely, however, you will notice that Transhumanism ushers in a bunch of quietly whispered "hopes" through the back door:

"I hope that we can control our technologies."

"I hope that super-intelligence won't turn on us."

"I hope bad people never get ahold of these devices."

"I hope these technologies will help me feel less lonely."

And on and on. Pleading that bad things will not come to pass is not hope. Wishing that human beings would be good to one another when in the possession of enormous power is not hope. It is folly. The Christian Scriptures offer something different. The truth. The truth about God's love for us and the truth about our condition as people. Our final chapter together will discuss who we are as humans in light of Scripture.

Human Uniqueness and the Trinity

What does it mean to be human? If you have made it this far, you deserve: (1) a medal, and (2) a well-grounded theological answer to this question. You have been waiting long enough!

I offer you a way forward in this chapter. My thoughts here coalesce many of the insights from previous chapters into one simple framework that I have found to be helpful in my own thinking. But it's only a framework. You might add to it. Amend it. At the very least, consider this a picture of Christian anthropology that merits discussion in your homes, parishes, and schools. Remember that every context has specific needs, and while I speak more generally about the characteristics we all share, spend time considering how what follows might be put to best use in your corner of the world.

I argue that the uniqueness of human identity is built on three pillars: vocation, embodiment, and church-community. These features loosely coincide with a creedal format where I link each one to a person of the Trinity.

Vocation

For a quick moment, return and skim chapter 7. There, I briefly introduced the concept of vocation and explained why it sits at the core of the human enterprise. It's time to go deeper. To have a *vocation* is to have a purpose. Vocation is the great engine of human resourcefulness, as it narrows the field of potential things a person *could* do (or be) down to the things that person *has been called* to do (or be). Once a person knows their role and calling, they direct their intellectual and physical energy to fulfill that potential.

Martin Luther, the great Reformer, understood the power of vocation as he observed the social structures of sixteenth-century Germany. He noticed a false dichotomy where the affairs of the world were commonly divided into two categories: either something was sacred (i.e., linked with the divine), or it was considered secular (i.e., worldly). This approach bothered Luther because it implied that only certain stations in life were able to please God. Everything that was not explicitly churchy was given a lower status. If you were a peasant in the sixteenth century, you would have been able to see quite clearly which jobs were "more godly" than others. Bishops, priests, and monks had special garments, special abilities, even special incantations. It was as if this class of clergy had access to the magical arts; they dealt with transcendence and mystery. How could the weaver possibly compete with that? What could be godly about twine and dye?

Luther thought this distinction was a mistake. After all, where in Scripture does it command a person to give higher status to a monk than a mother when all were a part of the same Body of Christ? His concept of vocation leveled the playing field. Now, all manner of work could be dignified as "work for the Lord" as long as it was connected to the faith of the doer. For example, Luther praises fathers and mothers who are blessed with the vocation of raising children and all the mess that parenthood entails. He scolds those who might suggest the mundane is without glory by saying,

> Now you tell me, when a father goes ahead and washes diapers or performs some other mean task for his child, and someone ridicules him as an effeminate fool—though that father is acting in the spirit just described and in Christian faith—my dear fellow you tell me, which of the two is most keenly ridiculing the other? God, with all his angels and creatures, is smiling—not because that father is washing diapers, but because he is doing so in Christian faith. Those who sneer at him and see only the task but not the faith are ridiculing God with all his creatures, as the biggest fool on earth. Indeed, they are only ridiculing themselves; with all their cleverness they are nothing but devil's fools.[1]

Dealing with diapers is not glorious or particularly noteworthy—except to the child. The Reformation fathers hit this one out of the park. All of one's life can be dedicated to the advancement of the gospel, not by memorizing the Greek translation of Romans but by serving those whom God has given us to serve, including (and most especially) the child on the changing table. Parenthood is so closely linked with vocation that it will not surprise you that I link God the Father to this component of human uniqueness.

1. Luther, "Estate of Marriage," 40–41. Luther was not suggesting that only men who changed diapers were performing vocational duties. He speaks at length about the value of mothers and their work in this and other texts.

Vocation and the Father

I suggest developing a clear definition of vocation and work toward a proper reasoning of why I think this belongs as a part of the framework. A simple definition of vocation would be to have a purpose that aligns with God's will for his own glory and oriented to the benefit of one's neighbor. It is a calling to work diligently in any task as if working for the Lord himself (Col 3:23). Historically, the term has been directly linked with one's formal employment. A shipbuilder has the vocation of building ships; the nurse has the vocation of assisting those who need medical care. I would suggest that the term should be more broadly understood to encompass all of the activities we do (and roles we fulfill) that could give God glory. I am not just a professor. I also have the vocations of husband, father, son, and citizen—each with responsibilities and opportunities to serve my neighbor in faith.

More than a consideration of roles, vocation also speaks to our purpose. *Why* do we fulfill those roles? I am a father, sure, but this is more than a designation of my responsibility to my children. Laden within that role is the divine purpose that I have. Specifically, my *telos* as a father is to raise my children with discipline and love in order that they might become disciples of the Lord Jesus. Father is more than a label. It is a clue that indicates a particular future that I am partnering with God to bring about.

God the Father expressed a desired future of his own the moment he brought the world into existence. In the Genesis creation narrative, he brings forth Adam and Eve from the ground to accomplish certain things at the outright. God calls them into their assigned vocations, just as a father might lovingly give his own child a task or assignment that would serve the family. Adam and Eve had several callings at once.

They were to be fruitful and multiply (Gen 1:28).

They were to name the animals (Gen 2:19).

They were to have dominion over the created world (Gen 1:28b).

They were to care for and tend the Garden (Gen 2:15).

And, of course, they were to obey God's commands as they did all of that (Gen 2:16–17).

In short, Adam and Eve were given tasks to accomplish by which they could honor God.

In Ephesians, Paul follows his famous "For it is by grace that we have been saved" dictum with an important follow-up verse: "For we are God's handiwork, created in Christ Jesus to do good works which God prepared in advance for us to do" (Eph 2:10). The verse essentially means that God has been preparing us to succeed all along; he has been laying down the proper foundations so that we might fulfill our vocations not in an accidental way but with his grand design in mind. Just as a father helps his son get an education by first earning a wage to afford a good college, our heavenly Father pays any cost—heals any wound—so that we might do what we were born to do: serve him with gladness!

It's Complicated

Vocation, if done right, is a powerful way to understand the various responsibilities a Christian bears every day. But things can get complicated when your job offers little to the world and even less to your personal well-being. I am going to offer to you two insights into the power (and complexity) of vocation.

You might be in a job right now that stimulates you. You look forward to work because you know you are skilled at it, and you know that your skills are being put to good use. If that's your case, I delight with you! Of course, many of you dread going to work. Reading a chapter about vocation feels like fingernails scratching on a chalkboard. Perhaps your boss is cruel. Working for him feels like exploitation, and therefore, to call your work a vocation feels hollow. Even worse, it feels like a form of manipulation, a quasi-Christian way to keep you under someone's boot without complaining. Or, maybe your current job isn't bad or degrading, it's just . . . well, *there*. You don't feel like your gifts are being used in any significant way. You don't feel like you are contributing anything valuable to society. Instead, work is drudgery.

I hear you. I don't make light of the pain these situations cause. Because sin separates us from God, his call is often hard to hear. Most of us have never had a moment like the prophets in the Old Testament had when they heard the direct, unmistakable voice of God telling them what they were to do in no uncertain terms. And yet, we are human. We have agency to work toward a life of vocation no matter how far off it feels at the moment.

I say this to remind you of an oft-forgotten component of vocation. Vocation is highly relational and largely benefits from others who understand their vocations properly, as well. When speaking of traditional forms of vocation like work, a contract of sorts is at play. Both the employee *and the employer* commit together to a job that best suits the worker's abilities, desires, and needs. The best working environments often emerge when the boss takes interest in your particular abilities (and aspirations) and puts them to direct use in ways that are edifying to the ecology of the company. What I am trying to dispel is the claim that vocation is a way for the powerful (the employer) to keep the relatively powerless (the employee) quiet and subservient. Rather, *both* employer *and* employee are to be about the business of fulfilling their *telos*, a purpose that works diligently to dignify all parties as they work together.

I would like to turn your attention to one other observation that is lurking in the shadows here. Counterintuitively, there exists a powerful affirmation of vocation that only reveals itself when a person is experiencing the pain of enduring a bad job or the suffering that comes with no job at all. When our jobs are unfulfilling or when we feel that we are outside of God's design for our purpose, the feeling is miserable. We are aimless. I would suggest that the person without vocation is highly susceptible to depression. This is the story of a person without a calling, without a way to serve the community, and such circumstances often lead the person to feel as if he is less than human. For me, unemployment is much more than a difficult period of financial hardship. I believe it invades the heart of a person's identity because it prompts questions like, "Why aren't my gifts being recognized?" "What do I have to offer the world anymore?" Unemployment threatens our sense of humanity.

I have seen many homemakers in my life who, once their last child moved off to college, experienced a sort of post-children life crisis. Some relished the chance to explore new opportunities for their life. Others got stuck in a dark place for some time. While I want to speak cautiously here, I am convinced that this darkness comes in large part from the removal of a long-time vocation. No longer a primary care-giver to her now-grown children, the mother must ask herself, "To what purpose shall I live for now?" Or even, "Who *am* I now that I'm not a caregiver?" In short, we fundamentally know how important healthy and fulfilling vocations are precisely because of the identity crises that are caused when we lose sight of them.

Identity is often bound in the various vocations we hold. Which is why I believe that we can see vocation in both active and passive terms.

Passively, I can view my current situation with eyes of thankfulness. If you have a job, be thankful for it knowing that many people suffer through unemployment. I find it a joyful experience to regularly stop and identify the other callings (or roles) that continue to bless my life in quiet, simple ways. Actively, I can pursue new avenues in my life that give God glory while also fighting against injustice. This pursuit can empower you to speak up for yourself when you are being taken advantage of or used in ways that are de-humanizing. Better yet, you have the ability to support your neighbors as they also fight for healthy expressions of vocation in their lives.

By God's grace (and knowing that this will not be fully realized until the Second Coming), communities of vocation will more closely reflect the new heavens and new earth with each member joyfully bearing in one another's burden in harmony with nature.

Unique in Purpose

You might be thinking right now that vocation could be applied to other parts of nature—not just human beings—particularly if you have a Christian understanding of God's good creation. After all, do not animals fulfill their God-given roles of animal-ness in nature? Aristotle, in many of his writings, tied the "goodness" of a creature to its given *telos*. In other words, a good bear is one that fulfills his purpose as a bear by doing things that a bear does! A bear that acts like a squirrel is a bad bear . . . quirkier no doubt, but bad nonetheless.

So why do I suggest that vocation (at least in the sense that I am using it) is uniquely human? Two reasons. While animals (in fact, all of creation) fulfill their purpose in God's creation, they are not given dominion. Dominion is not domination, though people often forget that fact. Dominion is more like stewardship, where Christians take on the responsibility of care for things that are not ultimately theirs. That's *everything*. Since God has entrusted nature to us, we are the guilty ones when things go bad . . . not rogue bears acting like squirrels. But dominion exercised properly allows all animals—in fact, all the natural world—to enjoy balance and interdependence in a way that praises God for his design. Only humans have the capacity to steward the created world.

Second, humans are unique from animals in that their vocation is central to the salvific narrative of the whole cosmos. Said more plainly, God's redemptive work happens through human beings, not aardvarks. We can think of this in a variety of ways. Negatively, only humans sin. They are the failed moral agents that need redemption. Therefore, the plan of salvation (the vocation of God, in one sense) is initiated to save all of humanity, and by extension, all of creation. Positively, Jesus gives the command to his disciples (also, not aardvarks) to be the bearers of the gospel in the Great Commission. Jesus himself—as human—is the vehicle by which all of the cosmos is redeemed and sanctified. Every Christian has at least this vocational role in common: to go make disciples, baptize, and teach.

Embodiment

Humans are *vocational* creatures. And yet they are most definitely vocational *creatures.* The second profound quality of human identity is our *embodiment.* By now, you know that I am "all in" when it comes to our wonderfully complex, wonderfully fragile bodies. To be human is to have a physical presence in real concrete environments; we are not simply minds housed in flesh-and-bone protective casing. We were never meant to be just consciousness. From time eternal, God designed us to be body and soul.

In my estimation, embodiment most clearly links with the second person of the Trinity. As you remember from chapter 4, incarnation is best described as a movement from image (or design) to flesh. Jesus Christ, the very Incarnation of God, is described in this way. "The Word became flesh and made his dwelling among us" (John 1:14). The idea of God becoming man is strange enough, setting Christians apart from all the other major religions from the outset. Yet even more mysterious is the fact that "in Christ all the fullness of the Deity lives in bodily form" (Col 2:9). Jesus was not an abridged or truncated version of God. He wasn't a cliff notes version of the book, so to speak. He was *fully* divine even when he was experiencing the all-too-human affairs of hunger, thirst, cold, abandonment, loneliness, joy, and grief. With this in mind, I would suggest that it is difficult to make the argument that our bodies are either faulty or (at best) unimportant, as our transhumanist friends might have us believe.

Jesus' body was crucified. He was raised *in the flesh.* And that has profound importance to human beings who must ask themselves why their own bodies matter.

God's Self-Communication to Us

I am sure there are a thousand reasons why embodiment is crucial to our identity as humans, but for the sake of brevity I'm going to offer just two. Both rely on the relational nature of being human. In one sense, God speaks to us through material means of which embodiment serves as the most clear example. In another sense, human-to-human communication flourishes when we do it face-to-face absent all the media that causes interference and misunderstanding. The former focuses on God's method of reaching us and the latter emphasizes our efforts to reach others.

First, our physical bodies matter because embodiment seems to be at the very heart of God's self-communication. He carries out his will by using people, materials, events . . . *things*. And by doing that, God has created a more tangible way for the believer to see and understand the divine mind, far better than discerning his plan through vague intuition or coincidence. While God is spirit (John 4:24), he works through material means to carry out his will throughout the Bible. Let me provide a few examples of what I am talking about. We are creatures of God's own hand. This simply means that our physicality is a part of God's purposeful design, whether we fully grasp the significance of that reality or not. We were built to have bodies! Our first relationship, therefore, is one of creature to Creator and *vice versa*. The embodiment of the individual reminds us that we our not unlimited or omnipotent like our Creator and that we must bear certain restrictions on our being in patience and humility. God's design includes partnering with other people—community, not just spouses—who might help us bear the burden of physical life.

Second, God has provided us with sacramental instances of self-disclosure, as well. In less painful terms, God gives us the sacraments (i.e., Holy Communion and Holy Baptism) by which we get spiritual benefit through material means. Every week, millions upon millions of Christians confess that the simple elements of water, wine, and bread communicate something much more than a gesture of faith. Recipients receive God's absolution, the very promise that Jesus' body and blood imparts life and forgiveness right now and the assurance that the Holy Spirit moves through the waters of baptism to confer faith on the baptized. God offers tangible grace for his church, bought with the tangible blood of Christ and offered in tangible elements for his people's ongoing growth.

God communicates himself in our created design and in the sacraments. To top it all off, Christians await the distinctly physical future promised by God in the last days. As I talked about at length in the last chapter, the biblical picture of the End Times is far from a view of heaven as a disembodied, spiritual place where souls float around on fluffy clouds. Instead, the author of Revelation speaks of the New Jerusalem in highly natural, overwhelmingly physical terms: rivers, crops, gates, foundations, gold, trees, light, and fruit! We were made as physical creatures, but God does not scrap his initial design once sin enters into the world. The resurrection of Jesus reminds us that God's *final* position is that our physical bodies are worth redeeming, and they are to be restored to their intended beauty, no longer subject to corruption or decay. It's as if God says, "You are going to be human like never before! Forever!"

I do not deny that there are times that God spoke to his prophets through dreams/visions, in ways that one would not describe as "embodied." Good! This truth can remind us that communication *can* happen through disembodied avenues, just as real communication can happen through Gmail. But this is the exception that proves the rule. Overwhelmingly, God reveals himself in theophanies (i.e., physical manifestations like the burning bush) and uses material means to impart his message of salvation in both testaments. Just like the sacrificial lambs in the Old Testament, Jesus shed actual blood from his actual body to impart actual forgiveness for those who would believe.

Our Communication to Others

God's self-communication is exemplified in the incarnation of Jesus. And yet, our bodies are crucially important for the reason that they act as instruments of communication *to one another* in ways that other manufactured media cannot reproduce.

Physical presence fosters the virtue of empathy. Empathy is not the capacity to say, "I know what you're going through." It's quite the opposite. It encourages the listener to build a relationship with his neighbor by starting from a position of ignorance. Empathy says, "I do *not* know what you are going through and I will not assume. Instead, I will listen to you, and by doing that, we can understand one another more clearly and build our relationship."

MIT professor and public intellectual Sherry Turkle has argued persuasively that face-to-face conversation is the cure to many of the problems we face in the Digital Age, from loneliness to the culture of outrage that fills social media. She begins with a simple yet profound observation: "Face-to-face conversation is the most human—and humanizing—thing we do. Fully present to one another, we learn to listen. It's where we develop the capacity for empathy. It's where we experience the joy of being heard, of being understood."[2] Face-to-face communication opens the door to the full gamut of human expression and largely reduces the chance that the two conversation partners misunderstand each other. The tilt of the head, the subtle arch of an eyebrow—each wink and gesture becomes a way by which I can know the mind and heart of my neighbor with better clarity.

The further we draw away from the face-to-face in favor of virtual or disembodied communication, the more likely we are to lose the parts of us that make us human. Without a doubt, our devices have made us less sensitive to the physical and emotional needs of those around us. Turkle suggests that "technology is implicated in an assault on empathy," and she is dead on.[3] Unfortunately, empathy is not the only casualty.

Directly related to empathy, physical presence develops intimacy. By using this term, I am inviting misinterpretation. Alas, I am writing my thoughts in a book and not having a face-to-face conversation with you, the reader! Intimacy can be generated in so many ways. I suggest, however, that our bodies serve as the best vehicles for communicating one's desire to be close to another, and I'm not talking about sexuality. Consider how underwhelming the "thumbs-up" icon in Facebook actually is. It makes use of a body part, excarnates it into an idea, and then tries to convey that idea as if it were the equivalent of a physical expression of affirmation. Now, consider the effect of a loved one's hand touching your arm gently, telling you without words that, "Ya did good."

Essentially, I am speaking more to relational closeness made available by physical proximity, a nearness that is fostered by seeing someone's full catalog of communication techniques. Sexuality *could* be included here, but intimacy itself (one fruit of the sexual experience) is a yearning to be more fully known in all of its intellectual and physical forms. Just as I want to be close with my Father in heaven, I am utterly disappointed that I don't have a body to hug and hold as the early disciples did to the resurrected Jesus. Hugging is knowing. The quest for intimacy is also why I give zerberts to my toddler. I want to bond with her physically and emotionally, and the physical act of the zerbert actually accomplishes both!

2. Turkle, *Reclaiming Conversation*, 3.

3. Turkle, *Reclaiming Conversation*, 4.

It is helpful to remember that physical closeness isn't just about times of warmth or friendship. Intimacy and empathy often make powerful partners in times of grief. One of my old colleagues once told me about a time when his daughter had to fight through a tragic loss and how that trauma grieved him as the father. He rhetorically asked, "What made me get in the car with my wife and drive fifteen hours to be with her in person?" He simply knew that calling, texting, even video-conferencing could not accomplish the type of solidarity that he needed in that moment. Only physical presence could do. On top of that, his decision to be with his daughter during this hard time forced him to partake in his daughter's sadness to a much higher degree. He could not simply "press send" and go back to his daily activities.

Finally, I suggest that embodiment is a much preferred, if unnecessary, component to the Christian call to witness. When Lebron James played basketball for the Cleveland Cavaliers in the mid-2000s, Nike initiated an ad campaign with the slogan, "We are all witnesses." The shoe giant was making a bold claim with that statement. In essence, they were suggesting that the basketball world was going to witness the next great (or perhaps greatest ever) basketball player emerge on the world stage. Prepare yourself to witness the amazing, they implied. Sure enough, Lebron's legacy is a powerful one. Those who talk about his feats on the hardwood share in the two-fold act I mention above. To properly defend his greatness, the person must first "witness" the event. They must see him on the court (or on the television) and catalog his deeds. But next, the act of witnessing includes a public speach about the events in question, so that the person's true testimony might influence a friend or neighbor who was not there. Witnessing for Christ, though it has become a pejorative in some social circles, can be reduced to that: (1) see what happened and (2) speak about what happened. When a person acts as a witness, they claim that their experience is a true and accurate picture of reality.

The Apostle Paul goes even further. Speaking to the church in Corinth, he suggests that the very bodies we inhabit function as a living testimony to the death and resurrection of Jesus. In 1 Cor 4, he writes:

> But we have this treasure in jars of clay to show that this all-surpassing power is from God and not from us. We are hard pressed on every side, but not crushed; perplexed, but not in despair; persecuted, but not abandoned; struck down, but not destroyed. We always carry around in our body the death of Jesus, so that the life

> of Jesus may also be revealed in our body. For we who are alive are always being given over to death for Jesus' sake, so that his life may be revealed in our mortal body. (2 Cor 4:7–11)

This is the testimony of martyrs, the story of how their bodies were given unto death in order that a greater claim could be made: Jesus is the Lord of life, death, and life again. The act of witnessing seems to be uniquely human after all. It is an expression of our bodily connection with the Son of God in his most horrible and most glorious moments.

The Community of Faith

The third and final piece of the puzzle is the church-community. To be human is to be in community with other people sharing the joys and pains of life together. The value of relationships is not restricted to Christians, of course; our species as a whole is unmistakably social! For believers, in particular, the community offers spiritual benefits that go far beyond the simple value of being less lonely. What follows is less about community as a sociological aid to foster human flourishing and more about the specific blessings that we receive as members of the Body of Christ. Therefore, I will focus my attention primarily on Christians in this chapter and use the term "church-community" as a way to tease out the social *and* spiritual dynamics of life together. But before we get to that, some important definitions are in order. The word "church" has different meanings in different contexts.

- Use #1: *Church*, with a big "C"

  Meaning: The eternal Church not bound by time or space. Sometimes this is called the "invisible" Church because it is not a concrete structure and it does not have an address. It is the universal fellowship of people who share belief/trust in the gospel of Jesus.

  Who's involved: All Christian believers since the beginning of time, including St. Augustine, Thomas Aquinas, John Calvin, Martin Luther King Jr., and your Methodist neighbor, Steve.

- Use #2: *church*, with a little "C"

  Meaning: The local, concrete parish in which Christians live together in real communities. These congregations have names like, "St. Paul's Lutheran" or "Word of Life Community Church." They usually find or build a place of worship and set about the tasks of witnessing to their neighborhoods and caring deeply for the spiritual needs of its members.

Who's involved: A local group of Christians who gather as one body to share a common profession of faith. They are often led by a pastor or priest who has multiple responsibilities, including the leadership of worship, administration of the church's duties to its community, and the pastoral care of its members.

My use of the word "church-community" will be restricted to the second definition, for it most closely reflects the concept I am trying to invoke. The church-community is made up of broken but forgiven people who love their neighbors; it is on the front lines of proclaiming the gospel to the surrounding culture. The reason I tack on the word, "community" to "church" is to remind me that the church is not about programming. Many people regard church as a place where Christians do Bible studies, retreats, potlucks, service projects, prayer vigils, women's groups, pancake fundraisers, and a thousand other things. These programs may be more or less important, but they are not essential to the life of the parish. The parish is made up of *people* who seek to follow Jesus. That is its core value. The church-community, therefore, is essentially social; it is a fellowship. Pancake dinners and community outreach events flow out of this core commitment to be, above all else, a fellowship of Christ followers.

Before I continue, however, I do want to offer one reason why we should always remember the importance of the *first use* of the word, Church. Even though the Church (big-C, see above) is not tangible so much as it is spiritual, it serves as the foundation for how we think about what it means to be Christian in relation to other Christians. A common Western approach is to think about a church as "something you attend" or "an organization that you join." And I suppose there is a measure of truth in this characterization, as many congregations require membership classes both to teach potential members the basic tenets of their church's core beliefs as well as weed out those who might be seeking church as a hobby rather than as a way of life. I would, however, caution us against thinking of church as a commodity to be bought and traded. The eternal Church of God was never something we could create or manufacture, as if a group of people came together one day and said, "You know what would be a great idea? Let's build something that will ensure that Christians have a place to be together!" The universal fellowship of believers is not something you *join*. The Church is a reality that you participate *in*.

Theologian Dietrich Bonhoeffer highlights this idea when he rejects human agency as the means by which we are included in the Church, as if joining the fellowship of God on earth was the same as joining a gym. This approach fosters an attitude of consumerism: I want to be a part of this church, I sign the documents and attend the proper membership classes, and I'm in (at least until the church displeases me and I move to another). By contrast, Bonhoeffer argues that Christian fellowship is not a thing to be fashioned but a reality to be received! "Christian brotherhood is not an ideal that we must realize; it is rather a reality created by God in Christ in which we may participate." He goes on to say, "because Christian community is founded solely on Jesus Christ, it is a spiritual and not a psychic reality. In this it differs absolutely from all other communities."[4] While many types of human associations have value, Bonhoeffer is pressing home the great advantage that exists for Christians: we have a life together founded in the person of Christ. Our fellowship is grounded in grace and truth (John 1:14) from the moment we enter into the community as baptized believers all the way into the perpetual future of eternal life, filled with joy and peace.

The church-community is not a commodity. While some congregations may be small is size, the promise of Jesus' presence guarantees that its work is not in vain. We can now ask ourselves: How is this gathering of kindred spirits distinguished from other meaningful associations? Why is it so important? How does it contribute to my identity as a created child of God?

The Church Family

It's no secret that God designed us to be in multiple forms of relationship: human-to-God, human-to-human, even human-to-nature. Evidence of this begins in Genesis 1 and continues to Revelation 22 (that's the whole bag, by the way). Other than Christology (i.e., everything in the Bible points to Jesus Christ), the presence and importance of these relationships might just be the most easily observed themes in all of Scripture. The Bible is essentially a story about a relationship gone bad and the unwavering commitment of God to mend the brokenness, making all things new again. Before that final renewal takes place, Christians can bring that future paradise to the present

4. Bonhoeffer, *Life Together*, 30–31.

in a small way by being a people who love each other, demonstrating to the world the new kingdom of God that Jesus initiated two-thousand years ago.

Regardless of a person's faith or lack thereof, the importance of social interaction cannot be denied even for the most introverted among us. We need communities to help us articulate a coherent understanding of the world and our place within it, to be sure, but we also need belonging and warmth that can only come from contact with people. A dog is man's best friend, but a dog can never fully understand the depth of human suffering, joy, emotion, sexual intimacy, and responsibility that comes with the territory of being made by God for the purposes of his glory. Our relationships with families and friends, though far from perfect, offer us a window into the mind of God. He has built structures of support into the very fabric of our species in order that we might know what it means to give and receive love in tangible ways, mirroring his own love as the Father.

The church-community offers something else, however, something distinctive that it alone is best designed to achieve. If I was strictly speaking of community from a sociological point of view, I would point to the family unit as the single most important association a person could have, responsible for developing identity and narrative for all of its members. The Christian recognizes the fundamental good of family and adds to it. In Matthew 12, Jesus is talking with his disciples when someone politely informs him that Jesus' family is outside wanting a word. Jesus responds in puzzling fashion, "Who is my mother, and who are my brothers?" And then, "Here are my mother and my brothers. For whoever does the will of my Father in heaven is my brother and sister and mother." Jesus' response is not meant to denigrate the family unit. Of course not. He is, however, recognizing a bond that exists at a bedrock level: the spiritual family of God. Those who believe together are knit together for eternity, regardless if they share the same DNA or not.

I admit that this is difficult for me to wrap my head around. My wife and children mean the world to me, and they know me better than anyone else on the planet—my joys, my dreams, and my struggles. Still, Jesus says what he says. Thankfully, being a child of God does not require us to disassociate ourselves from the people who matter most to us. Family is not a zero-sum game. However, the God-person relationship *transcends* all other relationships. We are, in the words of St. Paul, adopted sons and daughters of God, which implicitly means that you are now a part of a spiritual family placed here on earth but oriented toward heaven (Gal 4:5–7). My relationship with Jesus does not require that I end all other human-to-human commitments, family or otherwise. Rather, Jesus is the model and standard by which I understand how to live within those other relationships in accordance with God's design.

Jesus' call in the Matthew passage is an invitation to participate in a church-community of faith and trust, exceeding but not excluding the other significant relationships one can develop. He directs our gaze outward so that we might see the bigger picture of God's plan and also seek out those who yearn for the love of the Creator.

Needing the Outside Voice

Why is the church-community necessary? According to Jesus, the church-community is the family of God bonded together by love and bent toward the end of accomplishing his will. Assuming that its existence is God's design from the beginning, Christian fellowship must offer Christians something truly important, something *essential*. I would direct your attention to at least two reasons the church-community of God matters.

In the church-community, we receive the *external preaching of the gospel*. Yes, if a person is fortunate to grow up in a Christian family, they may hear the gospel often—even daily. To God be the glory! The church-community joyfully bears this responsibility to the world. They are entrusted with sharing the message of Jesus—the embodied life of faith in action—to all nations. Experiencing the proclaimed gospel "from the outside" is not to be dismissed as unimportant. Do you remember ever getting a compliment from a parent that you simply dismissed *because it was your parent who said it*? My brother once asked me to talk to his son, my nephew, about his future vocations. He said, "He's more likely to listen to you because it's not coming from his dad." There is power in hearing the gospel from someone on the outside, someone who is committed to preaching the objective Word of God. We need to hear it spoken to us, and we need to bear it to other people. The church-community is ground zero for the life of grace. We not only need to hear grace proclaimed to us (and sometimes discipline!) by our Christian brothers and sisters, but also that we ourselves must embrace the responsibility within that spiritual family to bear grace to others! Notice how far we are from a consumer approach to church with these words. Participating in the life of the church implies a humble willingness to offer the gospel to my neighbor who might desperately need God's grace in that very moment. And joy of all joys, you just might be the person who has the chance to offer it!

The church-community *receives the sacraments together*. The pastor or priest, as head of the local church, offers the sacraments to the congregants on behalf of Christ himself. Desperate times have often led heads of households to baptize their sons or daughters in isolation from their church communities, but this is only the exception to the rule. Baptisms offer forgiveness *and* community. The Holy Spirit works faith in the believer and calls together the entire Christian body to take ownership of the newly reborn child of God as one of their own. Baptism was never designed to be "a splash of water and a pat out the door"—as if it was a magic spell. Rather, the newly baptized believer was to be enfolded into an accountable family of faith where the benefits of water and Word could be fleshed out in a real community of support.

Similarly, the practice of Holy Communion is to be experienced within the context of the Christian family. The individual does not hoard the elements and retreat away from everyone else to partake in a private mystery of God's presence. The spiritual benefits of the meal play themselves out in the fellowship, where forgiveness of sins (an essential social experience) becomes the order of the day. Men and women receive the presence of Christ in the sacrament and then become Christ-like to one another as agents of grace. Each member receives the means of God's grace, then moves out into the world to be a bearer of the same, life-changing love.

The reality of the modern world is that our relatives are often far away. Many of us do not live in the same communities as our parents. Children eventually grow up and move off to different towns, even different states, as they follow their career path wherever it might lead. This is the long-term hardwired genius of Christian fellowship; it opens the definition of family to include our brothers and sisters in the faith and as such, we are never alone in our walk with the Lord.

You might say that the church-community is the walking, talking presence of the Holy Spirit. Just as we find identifying marks of the Father in vocation and features of the Son in embodiment, so the Holy Spirit is the living, breathing impetus for all that the church-community strives to be. The early Church was launched at Pentecost (Acts 2), a story so amazing (and funny) that it could only have been conceived by an uncontainable Holy Spirit who "blows wherever it pleases" (John 3:8). The Holy Spirit is the one who binds us in common faith to one another. The Holy Spirit brings faith to maturity in the hearts of the Church family. The Holy Spirit inspires us to broaden that family by giving us the courage to witness to a world largely unfamiliar with the ways of Jesus. It is the Holy Spirit who advocates on our behalf when our prayers fall short (Rom 8:26), when our energy becomes weak, and when our hearts feel heavy. In short, the Holy Spirit moves the Church to be the Church: a bold, confessing, forgiving, serving, worshiping, family of misfit believers bent on making the present world a little more like the world to come.

Much of what I've said in this chapter is theologically dense. I am a theologian and I'm prone to creating a complex mess of terms and concepts in my classroom. But this whole Church thing need not be so complicated. My wife has a saying that is as simple as it is profound. Whenever she witnesses her faith to a neighbor, her testimony often starts with the phrase, "Some girls at my church . . . " For example, if an acquaintance of my wife says,

- "Hey Tiffany, I heard you were sick last week. Everything okay?"

  She responds, "Yes! I'm great. *Some girls at my church* started a meal train and helped watch the kids during the day."

- "Hey Tiffany, how did you make it through that dark time in your life?"

  "*Some girls at my church* started a prayer chain and took turns visiting me at my home. I felt so honored and loved . . . "

- "Hey Tiffany, you're particularly chipper today! Why are you smiling?"

  "*Some girls at my church* brought me flowers and then stayed for lunch."

"Some girls at my church" is just another way to say that the Body of Christ (the church-community) did what it was designed to do. Serve, absolve, commune, baptize, lead, welcome, feed, care, and worship. These are the eternally important tasks of the church fellowship, and they are not impossible. The Holy Spirit, after all, gave the early disciples the ability to speak the Word of God to strangers in languages formerly unfamiliar to them. The same Spirit transformed that motley group of fearful fisherman into giants of the gospel. I'm sure that the Holy Spirit can handle your community of faith with its zits and scars, too.

Being Human

Vocation, embodiment, church-community. These are the qualities that make humans, human. But you need all three. To have a vocation and a solid sense of my physical embodiment but no community is to have a loner's existence with little spiritual support and even less accountability. Christians are not called to be hermits. To be embodied and connected to a community may have terrific benefit, but without vocation a person wanders in search of a purpose by which he can serve God in his community. Christians are not called to be freeloaders. To have a vocation and enjoy the life of the church-community and yet exist heavily in the virtual or digital realms is to forfeit the empathy, sexuality, care, and communication that can only come with a physical encounter with others. Christians are not called to be disembodied minds. We are designed to be the whole package, rich in complexity and worth far more than the sum of our parts.

The technological age can be difficult to navigate. The world changes so fast these days that every morning seems to present us with a new challenge to overcome. Thankfully, Jesus has overcome the world (John 16:33)! If this is true, then the world is not so much an enemy to be resisted with every fiber of our being, but an opportunity to be about the business of holy influence. When we watch the cultural tides ebb and flow with clear eyes, we can see both the world's resistance to God's design *and* God's movement to redeem it one bit at a time.

The Christian need not worry about this or that technological innovation. They merely need to evaluate it in the light of their own created design. Does this technology promote a sense of purpose and vocation? Does it dull my desire to contribute to my society in tangible ways? Does this innovation seek to improve or encourage my embodied presence in my community? Or, does it imply that I have no actual use for my flesh and bones? Does Transhumanism draw me in closer relationship to my community, or does it encourage me to think of myself as a self-contained, self-sustaining demigod? We don't have to be Luddites. We just have to have our eyes open. When we think intentionally about the issues of the Digital Age, the fear of technology begins to dissolve, leaving behind a simple (but perhaps, radical) commitment to a discerning life lived in joy.

Discussion Questions

1) What would you add to the three-part framework (vocation, embodiment, church-community) that the author suggests? Anything?
2) Of the three parts, which do you find resonates most with your view of human uniqueness?
3) Evaluate the short-comings of a transhumanist worldview through the lens of this chapter.

Bibliography

Bauerlein, Mark. *The Dumbest Generation: How the Digital Age Stupefies Young Americans and Jeopardizes Our Future*. New York: Penguin, 2008.

Bonhoeffer, Dietrich. *Letters and Papers from Prison*. New York: Touchstone, 1997.

———. *Life Together*. Translated by John Doberstein. New York: Harper & Row, 1954.

Bostrom, Nick. *Superintelligence: Paths, Dangers, Strategies*. Oxford: Oxford University Press, 2014.

Bulman, May. "EU to Vote on Declaring Robots to Be 'Electronic Persons.'" *Independent*, January 14, 2017. Accessed February 26, 2018. http://www.independent.co.uk/life-style/gadgets-and-tech/robots-eu-vote-electronic-persons-european-union-ai-artificial-intelligence-a7527106.html.

Burdett, Michael S. "The Religion of Technology: Transhumanism and the Myth of Progress." In *Religion and Transhumanism*, edited by Calvin Mercer and Tracy J. Trothen, 131–47. Santa Barbara, CA: Praeger, 2015.

Carr, Nicholas. *The Shallows: What the Internet Is Doing to Our Brains*. New York: Norton, 2011.

Castronova, Edward. *Exodus to the Virtual World: How Online Fun Is Changing Reality*. New York: Palgrave MacMillan, 2007.

Culkin, John M. "A Schoolman's Guide to Marshall McLuhan." *The Saturday Review* (March 1967) 51–53, 70–72.

De Grey, Aubrey. "The Curate's Egg of Anti-Anti Aging Bioethics." In *The Transhumanist Reader: Classical and Contemporary Essays on the Science, Technology, and Philosophy of the Human Future*, edited by Max More and Natasha Vita-More, 215–19. Oxford: Wiley-Blackwell, 2013.

Detweiler, Craig. *iGods: How Technology Shapes Our Spiritual and Social Lives*. Grand Rapids: Brazos, 2013.

Fukuyama, Francis. *Our Posthuman Future: Consequences of the Biotechnology Revolution*. New York: Picador, 2002.

Gardner, William. *Frames of Mind: The Theory of Multiple Intelligences*. New York: Basic, 2004.

Green, Green. *Bonhoeffer: A Theology of Sociality*. Grand Rapids, MI: Eerdmans, 1999.

Hart, David Bentley. *The Experience of God: Being, Consciousness, Bliss*. New Haven, CT: Yale University Press, 2013.

Hugo, Victor. *The History of a Crime*, Vol. 1. Boston: Little, Brown and Company, 1909.

Jenson, Robert. *On Thinking the Human: Resolutions of Difficult Notions*. Grand Rapids, MI: Eerdmans, 2003.

Levy, David. *Love and Sex with Robots: The Evolution of Human-Robot Relationships*. New York: HarperCollins, 2007.

Linhorst, Michael. "Could a Robot Be President?" *Politico*, July 8, 2017. Accessed March 16, 2018. https://www.politico.com/magazine/story/2017/07/08/robot-president-215342.

Luther, Martin. "The Estate of Marriage." In *Luther's Works*, Vol. 45: *Christian in Society II*, edited by Walther I. Brandt, 11–49. St. Louis: Concordia, 1962.

Mallinson, Jeff. *Sexy: The Quest for Erotic Virtue in Perplexing Times*. Irvine, CA: New Reformation, 2017.

McGonigal, Jane. *Reality Is Broken: Why Games Make Us Better and How They Can Change the World*. New York: Penguin, 2011.

Midgley, Mary. *The Myths We Live By*. New York: Routledge, 2004.

More, Max. "The Philosophy of Transhumanism." In *The Transhumanist Reader: Classical and Contemporary Essays on the Science, Technology, and Philosophy of the Human Future*, edited by Max More and Natasha Vita-More, 3–17. Oxford: Wiley-Blackwell, 2013.

Oesch, Joel. *More Than a Pretty Face: Using Embodied Lutheran Theology to Evaluate Community Building in Online Social Networks*. Eugene, OR: Wipf & Stock, 2017.

Paul, John, II. *The Gospel of Life: Evangelium Vitae*. Boston: Pauline, 1995.

Pearce, David. "The Hedonistic Imperative." Accessed October 1, 2020. https://www.hedweb.hedab.htm.

Postman, Neil. *Amusing Ourselves to Death: Public Discourse in the Age of Show Business*. New York: Penguin, 2005.

Putnam, Robert. *Bowling Alone: The Collapse and Revival of American Community*. New York: Simon & Schuster, 2000.

Redding, Micah. "Christian Transhumanism Is the Next Reformation." *HuffPost*, October 17, 2017. Accessed September 6, 2018. https://www.huffingtonpost.com/entry/christian-transhumanism-is-the-next-reformation_us_59e40d37e4b09e31db975a6c.

Rice, Jesse. *The Church of Facebook: How the Hyperconnected Are Redefining Community*. Colorado Springs: Cook, 2009.

Rodriquez, Cecilia. "Sex-Dolls Brothel Opens in Spain and Many Predict Sex-Robots Tourism Soon to Follow." *Forbes*, Feburary 28, 2017. Accessed September 24, 2019. https://www.forbes.com/sites/ceciliarodriguez/2017/02/28/sex-dolls-brothel-opens-in-spain-and-many-predict-sex-robots-tourism-soon-to-follow/#61984fe24ece.

Ruse, Michael. *Monad to Man: The Concept of Progress in Evolutionary Biology*. Cambridge: Harvard University Press, 1996.

Sandberg, Anders. "Morphological Freedom—Why We Not Just Want It, but Need It." In *The Transhumanist Reader: Classical and Contemporary Essays on the Science, Technology, and Philosophy of the Human Future*, edited by Max More and Natasha Vita-More, 56–64. Oxford: Wiley-Blackwell, 2013.

Shirky, Clay. *Cognitive Surplus: How Technology Makes Consumers into Collaborators*. New York: Penguin, 2010.

Taylor, Charles. *A Secular Age*. Cambridge, MA: Harvard University Press, 2007.

"The Transhumanist Declaration (2012)." In *The Transhumanist Reader: Classical and Contemporary Essays on the Science, Technology, and Philosophy of the Human Future*, edited by Max More and Natasha Vita-More, 54–55. Oxford: Wiley-Blackwell, 2013.

Turkle, Sherry. *Life on the Screen: Identity in the Age of the Internet.* New York: Touchstone, 1995.

———. *Reclaiming Conversation: The Power of Talk in a Digital Age.* New York: Penguin, 2015.

Wakefield, Jane. "MEPs vote on robots' legal status—and if a kill switch is required." *BBC News*, January 12, 2017. Accessed March 12, 2018. https://www.bbc.com/news/technology-38583360.

Wright, N. T. *Surprised by Hope: Rethinking Heaven, the Resurrection, and the Mission of the Church.* New York: HarperCollins, 2009.

Zeigler, Michael. *Christian Hope among Rivals.* Eugene, OR: Wipf & Stock, 2017.

Printed in the USA
CPSIA information can be obtained
at www.ICGtesting.com
LVHW011043131223
766398LV00035B/651

9 781725 28732